HOLT, RINEHART AND WINSTON

Expressions and Formulas

BRITANNICA Mathematics in Context

Britannica

Mathematics in Context is a comprehensive curriculum for the middle grades. It was developed in collaboration with the Wisconsin Center for Education Research, School of Education, University of Wisconsin–Madison and the Freudenthal Institute at the University of Utrecht, The Netherlands, with the support of National Science Foundation Grant No. 9054928.

National Science Foundation
Opinions expressed are those of the authors
and not necessarily those of the Foundation

Revision Project

Peter Sickler
Project Director

Teri Hedges
Revision Consultant

Nieka Mamczak
Revision Consultant

Erin Turner
Revision Consultant

Cheryl Deese
MiC General Manager

Vicki Mirabile
Project Manager

2003 Printing by Holt, Rinehart and Winston
Copyright © 2003
Encyclopædia Britannica, Inc.

All rights reserved.
Printed in the United States of America.

This work is protected under current U.S. copyright laws, and the performance, display, and other applicable uses of it are governed by those laws. Any uses not in conformity with the U.S. copyright statute are prohibited without our express written permission, including but not limited to duplication, adaptation, and transmission by television or other devices or processes. For more information regarding a license, write Encyclopædia Britannica, Inc., 310 South Michigan Avenue, Chicago, Illinois 60604.

ISBN 0-03-071444-3

8 9 10 11 018 11 10 09 08 07

The *Mathematics in Context* Development Team

Mathematics in Context is a comprehensive curriculum for the middle grades. The National Science Foundation funded the National Center for Research in Mathematical Sciences Education at the University of Wisconsin–Madison to develop and field-test the materials from 1991 through 1996. The Freudenthal Institute at the University of Utrecht in The Netherlands, as a subcontractor, collaborated with the University of Wisconsin–Madison on the development of the curriculum.

The initial version of *Expressions and Formulas* was developed by Koeno Gravemeijer, Anton Roodhardt, and Monica Wijers. It was adapted for use in American schools by Beth R. Cole and Gail Burrill.

National Center for Research in Mathematical Sciences Education Staff

Thomas A. Romberg
Director

Joan Daniels Pedro
Assistant to the Director

Gail Burrill
Coordinator
Field Test Materials

Margaret R. Meyer
Coordinator
Pilot Test Materials

Mary Ann Fix
Editorial Coordinator

Sherian Foster
Editorial Coordinator

James A. Middleton
Pilot Test Coordinator

Project Staff

Jonathan Brendefur
Laura J. Brinker
James Browne
Jack Burrill
Rose Byrd
Peter Christiansen
Barbara Clarke
Doug Clarke
Beth R. Cole

Fae Dremock
Jasmina Milinkovic
Margaret A. Pligge
Mary C. Shafer
Julia A. Shew
Aaron N. Simon
Marvin Smith
Stephanie Z. Smith
Mary S. Spence

Freudenthal Institute Staff

Jan de Lange
Director

Els Feijs
Coordinator

Martin van Reeuwijk
Coordinator

Project Staff

Mieke Abels
Nina Boswinkel
Frans van Galen
Koeno Gravemeijer
Marja van den Heuvel-Panhuizen
Jan Auke de Jong
Vincent Jonker
Ronald Keijzer

Martin Kindt
Jansie Niehaus
Nanda Querelle
Anton Roodhardt
Leen Streefland
Adri Treffers
Monica Wijers
Astrid de Wild

Table of Contents

Letter to the Student vi

Section A **Arrow Language**
Bus Riddle 1
Airline Reservations 4
Wandering Island 5
Fish 6
Summary 6

Section B **Smart Calculations**
Making Change 7
Skillful Computations 11
Summary 14

Section C **Formulas**
Supermarket 15
Taxi Fares 17
Stacking Cups 19
Stacking Chairs 21
Bike Sizes 22
Are They the Same? 24
Summary 25

Section D **Reverse Operations**
Foreign Money 27
Going Backwards 30
Beech Trees 31
Ham and Cheese 32
Summary 32

Section E **Tables**
Apple Crisp 33
Shuttle Service 34
Home Repairs 38
Summary 40

Section F **Order of Operations**
Arithmetic Trees 41
Flexible Computation 47
Return to the Supermarket 49
What Comes First? 51
Summary 52

Try This! 54

Mathematics in Context • Expressions and Formulas

Dear Student,

Welcome to *Expressions and Formulas*.

Have you ever wondered how to figure out what size bike you need? There is a formula that uses the length of your legs to tell you. In this unit, you will see this and many other formulas. Sometimes you will figure out the formula by yourself. Sometimes the formula will be given to you, and you will use it to answer questions.

In this unit, you will also learn new forms of mathematical writing. You will use "arrow language," arithmetic trees, and parentheses. These different ways of writing will help you understand and use formulas and other math expressions.

As you work on this unit, you may want to keep an eye out for formulas that you see outside of math class. They are everywhere!

Sincerely,

The Mathematics in Context Development Team

A. ARROW LANGUAGE

Bus Riddle

Pretend that you are a bus driver. Early one morning, you leave the garage with no passengers. At the first stop, 10 people get on the bus. At the next stop, six people get on. At the next stop, four people get off the bus, and seven get on. After that, five people get on, and two people get off. At the next stop, four people get off, and no one gets on.

1. How old is the bus driver?

2. Did you expect the first question to be about the number of passengers on the bus?

3. How could you determine the number of passengers on the bus after the last stop mentioned above?

Mathematics in Context • Expressions and Formulas

Expressions and Formulas

When four people got off the bus and seven got on, the number of people on the bus changed. There were three more people on the bus.

4. Here is a record of people getting on and off the bus for another part of the day. Copy the chart into your notebook. Then fill in the missing numbers.

Number of Passengers Getting off the Bus	Number of Passengers Getting on the Bus	Change
5	8	3 more
9	13	
16	16	
15	8	
9	3	
		5 fewer

5. Look at the last row in the chart. What can you say about the numbers of passengers getting on and off the bus when you know only that there are five fewer people on the bus?

For the story on page 1, you might have kept track of the number of passengers on the bus by writing:

$$10 + 6 = 16 + 3 = 19 + 3 = 22 - 4 = 18$$

6. Do you think that writing the numbers in this way is acceptable in mathematics? Why or why not?

To avoid problems with the equals sign (=), you could write the calculation like this:

$$10 \xrightarrow{+6} 16 \xrightarrow{+3} 19 \xrightarrow{+3} 22 \xrightarrow{-4} 18$$

Each change is shown with an arrow.

This way of writing a string of calculations is called *arrow language*. You can use arrow language to describe any sequence of additions and subtractions, whether it is about passengers, money, or anything else.

7. Why is arrow language a better way to keep track of a changing total?

A. Arrow Language

Ms. Moss had $1,235 in her bank account. Then she took out $357. Two days later, she withdrew $275 from her account.

8. Use arrow language to show what happened and how much money she had in her account at the end of the story.

John had $37 before he earned $10 for delivering newspapers one Monday. The same day, he spent $2 for an ice-cream cone. Tuesday, he visited his grandmother and earned $5 for washing her car. Wednesday, he earned $5 for baby-sitting. On Friday, he spent $2.75 for a hamburger and fries and $3.00 for a magazine.

9. a. Use arrow language to show how much money John had left.

  b. Suppose John wanted to buy a radio that cost $53. Did John have enough money to buy it at any time during the week described above? If so, when?

In a region known for having lots of snow in the winter, there were 42 inches of snow on the ground one Sunday. This was what happened during the following week:

Monday	It snowed 20.25 inches.
Tuesday	It warmed up, and 8.5 inches of snow melted.
Wednesday	Two inches of snow melted.
Thursday	It snowed 14.5 inches.
Friday	It snowed 11.5 inches in the morning and then stopped.

10. How deep was the snow on Friday afternoon?

Mathematics in Context • Expressions and Formulas

Expressions and Formulas

Airline Reservations

There are 375 seats on a flight to Atlanta, Georgia that will leave on March 16th. On March 11th, 233 of the seats have been reserved. The airline continues to take reservations and cancellations until the plane takes off. If there are more seats reserved than there are seats on the plane, the airline creates a waiting list.

The table below shows what happens over the next five days.

11. Copy and complete the table.

Date	Seats Requested	Cancellations	Total Seats Reserved
3/11			233
3/12	47	0	
3/13	51	1	
3/14	53	0	
3/15	5	12	
3/16	16	2	

12. Write one or more arrow strings to show the calculations you made to complete the table.

13. When does the airline need to form a waiting list?

14. Toni, a reservations agent, suggests that, instead of arrow strings, it would be easier to add all the new reservations and then subtract all of the cancellations to get the total number of reserved seats. What are the advantages and disadvantages of her suggestion?

A. Arrow Language

Wandering Island

Wandering Island constantly changes shape. One side washes away, while on the other side, sand washes onto shore. The islanders wonder whether their island is getting larger or smaller. In 1988, the area of the island was 210 square kilometers. Since then, the islanders have kept track of the area that has washed away and the area that is added to the island.

Year	Area Washed Away (in km^2)	Area Added (in km^2)
1989	5.5	6.0
1990	6.0	3.5
1991	4.0	5.0
1992	6.5	7.5
1993	7.0	6.0

15. What was the area of the island at the end of 1991?

16. **a.** At the end of 1993, was the island larger or smaller than it was in 1988?

    b. Explain or show how you got your answer to part **a.**

Mathematics in Context • Expressions and Formulas

Expressions and Formulas

Fish

At the beginning of the year, Lake Mason held about 30,000 fish. During the year, 12,000 new fish hatched, and 5,000 others died naturally. Another 6,000 were caught by fishermen.

17. **a.** Write an arrow string to show the changes in the number of fish.

   b. What is the total change in the number of fish?

18. **a.** Describe the total change using only one arrow.

$$30{,}000 \xrightarrow{\;?\;} \underline{\quad}$$

   b. How did you decide what number to place over the arrow?

This string was written for another lake:

$$40{,}000 \xrightarrow{+10{,}000} \underline{\quad} \xrightarrow{+15{,}000} \underline{\quad} \xrightarrow{-8{,}000} \underline{\quad} \xrightarrow{\quad} \underline{\quad}$$

19. What change should go above the last arrow so that the final number of fish is the same as the beginning number? Explain how you decided.

Summary

Arrow language can be helpful.

A calculation can be described with an arrow:

$$\textbf{starting number} \xrightarrow{\text{action}} \textbf{resulting number}$$

A series of calculations can be described by a string of arrows:

$$10 \xrightarrow{+6} 16 \xrightarrow{+3} 19 \xrightarrow{+3} 22 \xrightarrow{-4} 18$$

Summary Questions

20. **a.** Find the result of the following arrow string:

$$12.30 \xrightarrow{+1.40} \underline{\quad} \xrightarrow{-0.62} \underline{\quad} \xrightarrow{+5.83} \underline{\quad} \xrightarrow{-1.40} \underline{\quad}$$

   b. Make up a story that could go with the arrow string in part **a.**

21. Make up a problem that you could solve using arrow language. Then solve the problem.

22. Why is arrow language useful?

B. SMART CALCULATIONS

Making Change

Today, making change in a store is easy. You just push some buttons, and the cash register shows the amount of change due. Before computerized cash registers, however, making change was not quite so easy. Because of this, people developed a strategy for making change, which is still useful in places where computerized registers are not available.

1. a. When you buy something, how do you know if you were given the correct change?

b. If you had to make change without a cash register, how would you do it?

Think about the following:

A total purchase is $3.70. The customer gives the clerk a $20 bill.

2. Explain how to figure the correct change without using pencil and paper or a calculator.

Mathematics in Context • Expressions and Formulas

Expressions and Formulas

It would be useful to have a strategy that works in any situation. Rachel suggests beginning by estimating. "In the example," she explains, "it is easy to tell that the change will be more than $15." She says that the first step is to give the customer $15.

Rachel explains that once the $15 is given, it is as if the customer has paid only $5. "Now, only the difference between $5.00 and $3.70 must be found, and that is $1.30, or one dollar, one quarter, and one nickel."

3. Do you think that Rachel has come up with a good strategy? Why or why not?

Many people use a slightly different strategy that pays small coins and bills first. Remember, the total cost is $3.70, and the customer is paying with a $20 bill.

The clerk first gives a nickel and says, "$3.75."

Next, the clerk gives a quarter and says, "That's $4."

Then, the clerk gives a dollar and says, "That's $5."

The clerk then gives a $5 bill and says, "That makes $10."

Finally, the clerk gives a $10 bill and says, "That makes $20."

Someone thinks that this method should be called "making change to five dollars."

B. Smart Calculations

One person suggests that this method uses the fewest coins and bills. Someone else says that this could be called the "small-coins-and-bills-first method."

4. a. Does this method, in fact, use the fewest possible coins and bills?

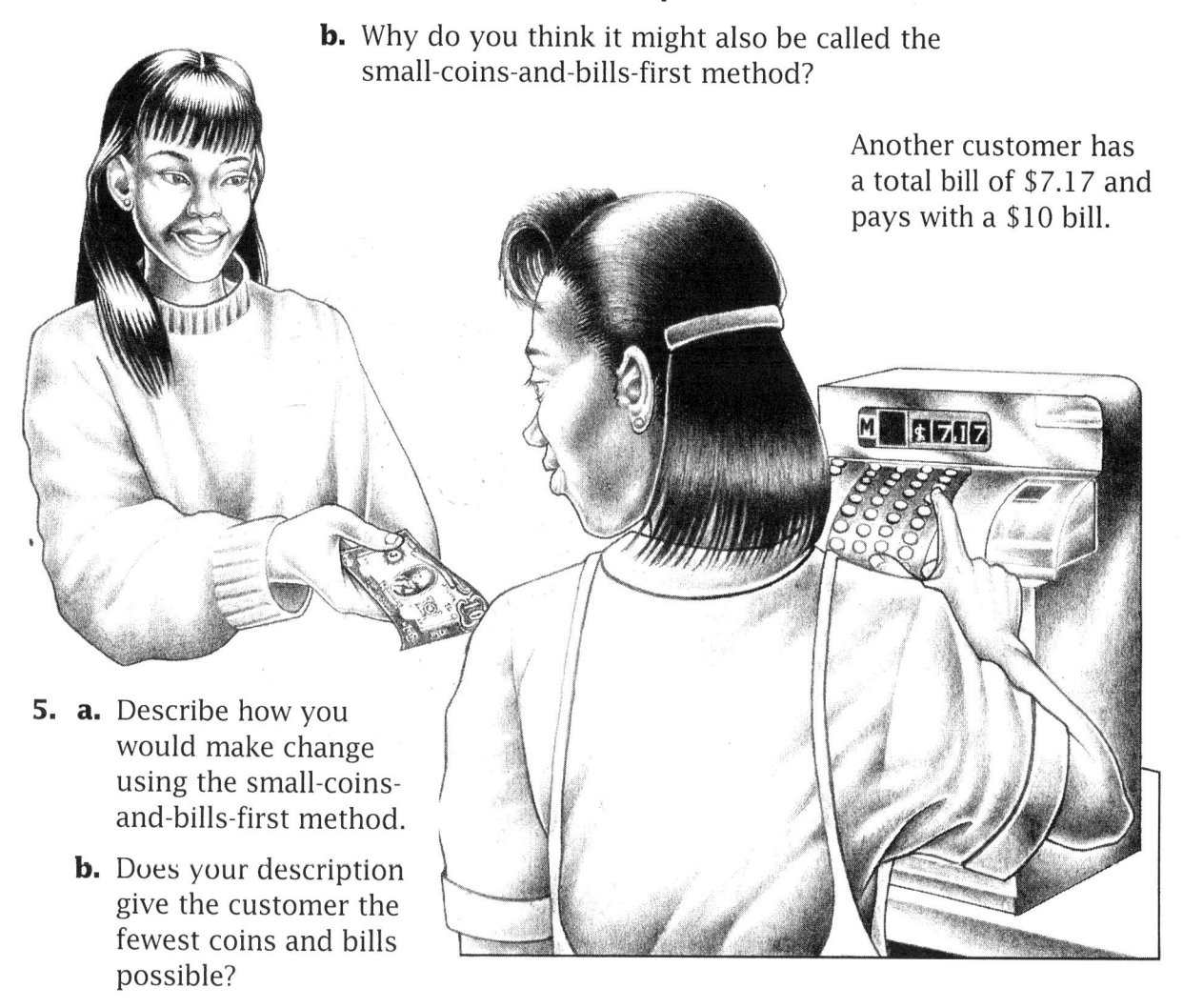

   b. Why do you think it might also be called the small-coins-and-bills-first method?

Another customer has a total bill of $7.17 and pays with a $10 bill.

5. a. Describe how you would make change using the small-coins-and-bills-first method.

   b. Does your description give the customer the fewest coins and bills possible?

One problem with the small-coins-and-bills-first method is that you do not know the total change. Arrow language can be used to show this method of making change. Here is how the clerk could show the change for the $3.70 purchase.

$$\$3.70 \xrightarrow{+\$0.05} \$3.75 \xrightarrow{+\$0.25} \$4.00 \xrightarrow{+\$1.00} \$5.00 \xrightarrow{+\$5.00} \$10.00 \xrightarrow{+\$10.00} \$20.00$$

6. a. Write what the cashier would say for the third arrow above.

   b. What is the total amount of change?

   c. Write a new arrow string with the same beginning and end but with only one arrow. Explain your reasoning.

Mathematics in Context • Expressions and Formulas

Expressions and Formulas

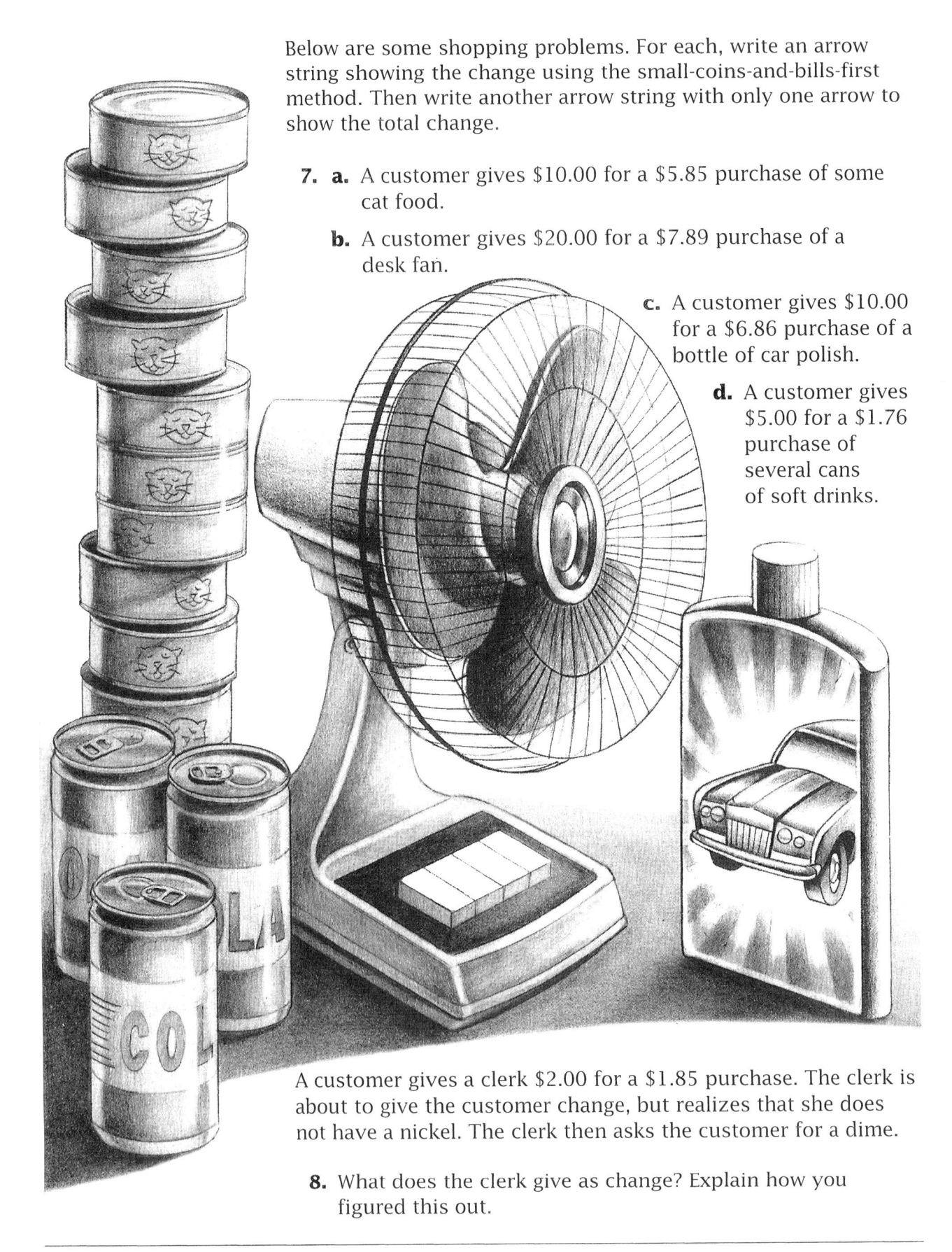

Below are some shopping problems. For each, write an arrow string showing the change using the small-coins-and-bills-first method. Then write another arrow string with only one arrow to show the total change.

7. a. A customer gives $10.00 for a $5.85 purchase of some cat food.

b. A customer gives $20.00 for a $7.89 purchase of a desk fan.

c. A customer gives $10.00 for a $6.86 purchase of a bottle of car polish.

d. A customer gives $5.00 for a $1.76 purchase of several cans of soft drinks.

A customer gives a clerk $2.00 for a $1.85 purchase. The clerk is about to give the customer change, but realizes that she does not have a nickel. The clerk then asks the customer for a dime.

8. What does the clerk give as change? Explain how you figured this out.

B. Smart Calculations

Skillful Computations

In problem **7,** you wrote two arrow strings for the same problem. One had many arrows, and the other had only one arrow.

9. Shorten the following arrow strings so that each has only one arrow.

  a. $375 \xrightarrow{+50} \underline{\ ?\ } \xrightarrow{+50} \underline{\ ?\ }$ is the same as $375 \xrightarrow{\ ?\ } \underline{\ ?\ }$

  b. $158 \xrightarrow{-1} \underline{\ ?\ } \xrightarrow{+100} \underline{\ ?\ }$ is the same as $158 \xrightarrow{\ ?\ } \underline{\ ?\ }$

  c. $1{,}274 \xrightarrow{-1{,}000} \underline{\ ?\ } \xrightarrow{+2} \underline{\ ?\ }$ is the same as $1{,}274 \xrightarrow{\ ?\ } \underline{\ ?\ }$

Some of these arrow strings, such as **9b** above, are easier to use for calculation when they are long. Others, such as **9a** above, are easier to use when they are short. Sometimes an arrow string can be made easier to use by making it shorter or longer.

10. For each of the arrow strings below, make a longer string that is easier. Then use the new arrow string to find the result.

  a. $527 \xrightarrow{+99} \underline{\ ?\ }$

  b. $274 \xrightarrow{-98} \underline{\ ?\ }$

Mathematics in Context • Expressions and Formulas

Expressions and Formulas

11. Change each of the following calculations into an arrow string with one arrow. Then make a longer arrow string that is easier to use.

   a. 1,003 − 999

   b. 423 + 104

   c. 1,793 − 1,010

12. a. Guess the result of the arrow string below and then copy and complete it in your notebook.

$$273 \xrightarrow{-100} \underline{} \xrightarrow{+99} \underline{}$$

   b. If the 273 in part **a** is replaced by 500, what is the new result?

   c. What if 1,453 is used instead of 273?

   d. What if 76 is used?

   e. What if 112 is used?

   f. Use one arrow to show what happens no matter what the first number is.

B. Smart Calculations

You can break up a number in many ways, making it easier to perform calculations. To calculate 129 + 521, you can write 521 as 500 + 21 and use an arrow string.

$$129 \xrightarrow{+21} 150 \xrightarrow{+500} 650$$

Sarah computed 129 + 521 as:

$$129 \xrightarrow{+500} \underline{} \xrightarrow{+20} \underline{} \xrightarrow{+1} \underline{}$$

13. Is Sarah's method correct?

14. How could you rewrite 267 − 28 to make it easier to compute?

15. For each of the arrow strings below, either write a new string that will make the computation easier and explain why it is easier, or explain why the string is already the easiest it can be.

    a. $423 \xrightarrow{+237} \underline{}$

    b. $554 \xrightarrow{-24} \underline{}$

    c. $29 \xrightarrow{+54} \underline{} \xrightarrow{-25} \underline{}$

    d. $998 \xrightarrow{+34} \underline{}$

Mathematics in Context • Expressions and Formulas

Expressions and Formulas

Summary

Sometimes an arrow string can be replaced by a shorter one that is easier to use.

___ $\xrightarrow{+64}$ ___ $\xrightarrow{+36}$ ___ becomes ___ $\xrightarrow{+100}$ ___

Sometimes an arrow string can be replaced by a longer one to make the calculation easier without changing the result.

___ $\xrightarrow{-99}$ ___ becomes ___ $\xrightarrow{+1}$ ___ $\xrightarrow{-100}$ ___

Summary Questions

16. Write two examples in which a shorter string would be easier to use. Be sure to include both the short and long strings for each example.

17. Write two examples in which a longer string would be easier to use. Show both the short and long strings for each example.

18. Write an arrow string to show how you would make change for the following purchases:

    a. A $6.77 purchase for which the customer pays $10.00.

    b. A $12.20 purchase for which the customer pays $20.00.

19. Explain why knowing how to shorten an arrow string can be useful in making change.

C. FORMULAS

Supermarket

1. Tomatoes cost $1.50 a pound. Carl buys two pounds of tomatoes.

   a. Find the total price of Carl's tomatoes.

   b. Write an arrow string that shows how you found the price.

At Veggies-R-Us, you can weigh fruits and vegetables yourself and find out how much your purchase will cost. There are buttons on the scale to indicate what is being weighed.

The scale's built-in calculator computes the purchase price and prints out a small price tag. The price tag shows what is being bought, the price per pound, how many pounds are being purchased, and the total price.

The scale, like an arrow string, takes the weight as an input and gives the price as an output.

weight $\longrightarrow$ ☐ $\longrightarrow$ price

Mathematics in Context • Expressions and Formulas

Expressions and Formulas

2. Find the price for each of the following weights of tomatoes using the following arrow string:

$$\text{weight} \xrightarrow{\times \$1.50} \text{price}$$

   a. 4 lb $\xrightarrow{\times \$1.50}$? b. 3 lb $\xrightarrow{\times \$1.50}$?

   c. 0.5 lb $\xrightarrow{\times \$1.50}$? d. 2.5 lb $\xrightarrow{\times \$1.50}$?

The prices of other fruits and vegetables are calculated in the same way. Green beans cost $0.90 per pound.

3. Write an arrow string to show the calculation the scale would use for green beans.

The calculation for the price of grapes is given by this arrow string: $\text{weight} \xrightarrow{\times \$1.70} \text{price}$

4. Draw price tags for the following purchases:

   a. 2 lb of grapes b. 3 lb of green beans

   c. 6 lb of tomatoes

The Corner Store does not have a fancy scale. The price of tomatoes there is $1.20 per pound. Siu bought some tomatoes, and her bill was $6.

5. What was the weight of the tomatoes Siu bought? How did you calculate this?

The manager of the Corner Store wants customers to be able to estimate the prices of their purchases. She posts a chart next to a regular scale.

6. a. Help the manager by copying and completing the chart below.

Weight	Tomatoes $1.20/lb	Green Beans $0.80/lb	Grapes $1.90/lb
0.5 lb			
1.0 lb			
2.0 lb			
3.0 lb			

   b. The manager wants to add more rows to the chart. Add a row to your chart to show the prices for 2.5 pounds of each item.

   c. Add and complete at least three more rows to the chart.

C. Formulas

TAXI FARES

In some cabs, the price for the ride is shown on a meter. At the Rainbow Cab Company, the price increases during the ride depending on the distance traveled. There is a starting amount you have to pay no matter how far you go, as well as a price for each mile you ride. The Rainbow Cab Company charges the following rates:

$ 09.50

The starting amount is $2.00.
The amount per mile is $1.50.

7. What is the price for each of these rides?

   a. from the stadium to the railroad station: 4 miles
   b. from a suburb to the center of the city: 7 miles
   c. from the convention center to the airport: 20 miles

The meter has a built-in calculator to find the price of a ride. The way the meter works can be described with an arrow string.

8. Which of these strings will give the correct price? Explain your answer.

   number of miles $\xrightarrow{\times \$1.50}$ ____ $\xrightarrow{+ \$2.00}$ total price

   $2.00 $\xrightarrow{+ \text{ number of miles}}$ ____ $\xrightarrow{\times \$1.50}$ total price

   number of miles $\xrightarrow{+ \$2.00}$ ____ $\xrightarrow{\times \$1.50}$ total price

Mathematics in Context • Expressions and Formulas

Expressions and Formulas

The Rainbow Cab Company has changed its rates. The new prices can be found using the following string:

number of miles $\xrightarrow{\times \$1.30}$ ____ $\xrightarrow{+ \$3.00}$ total price

9. Is a cab ride now cheaper or more expensive than it was before?

10. Use the new rate to find the price for each of the following rides:

   a. from the stadium to the railroad station: 4 miles

   b. from a suburb to the center of the city: 7 miles

   c. from the convention center to the airport: 20 miles

   d. Compare your answers before the rate change (from problem **7**) with those after the rate change. Did you answer problem **9** correctly?

11. After the company changed its rates, George slept through his alarm and had to take a cab to work. He was surprised at the cost: $18.60!

   a. Use the new rate to find out how far it is from George's home to work.

   b. Write an arrow string to show your calculations for part **a**.

The arrow string for the price of a taxi ride tells how to find the price for any number of miles.

number of miles $\xrightarrow{\times \$1.30}$ ____ $\xrightarrow{+ \$3.00}$ total price

An arrow string that tells how to do a specific type of calculation is called a *formula*.

C. Formulas

Activity

Stacking Cups

Materials:

Each group will need a centimeter ruler and at least four cups that are the same. Plastic cups from sporting events or fast-food restaurants work well.

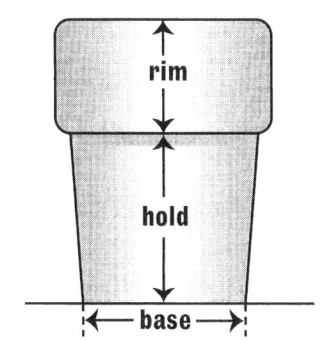

12. Measure and record the following:
- the total height of a cup
- the height of the rim
- the height of the hold

(*Note:* The *hold* is the distance from the bottom of the cup to the bottom of the rim.)

13. Stack two cups. Measure the height of the stack.

14. a. Without measuring, guess the height of a stack of four cups.

  b. Write down how you made your guess. With a partner, share your guess and how you arrived at it.

  c. Make a stack of four cups and measure it. Was your guess correct?

15. Use the method you discovered in problem **14** to calculate the height of a stack of 17 cups. Describe your calculation with an arrow string.

16. Choose two different numbers of cups to put in a stack. For each number, calculate the height. Then make a stack and check your calculation.

17. a. There is a space under a counter where cups will be stored. The space is 50 centimeters high. How many cups can be stacked to fit under the counter?

  b. Use arrow language to explain how you found the answer.

Mathematics in Context • Expressions and Formulas

Expressions and Formulas

Sometimes a formula is useful. You can write a formula to find the height of a stack of cups if you know the number of cups.

18. Complete the following arrow string for a formula using the number of cups as the input and the height of the stack as the output.

$$\text{number of cups} \xrightarrow{?} \underline{} \xrightarrow{?} \text{height of stack}$$

Suppose a class down the hall has *different* cups. The students use the following formula for finding the height of a stack of their cups:

$$\text{number of cups} \xrightarrow{-1} \underline{} \xrightarrow{\times 3} \underline{} \xrightarrow{+15} \text{height of stack}$$

19. a. How tall is a stack of 10 of these cups?

  b. How tall is a stack of five of these cups?

  c. Sketch one of the cups from this class. Label your drawing with the correct height.

  d. Explain what each of the numbers in the formula represents.

Now consider the following formula:

$$\text{number of cups} \xrightarrow{12 \times 3} \underline{36} \xrightarrow{+12} \underline{48} \text{height}$$

20. Could this formula be for the same cup from Problem **19**? Explain.

21. These cups also need to be stored in a space 50 centimeters high. How many of these cups can be placed in a stack? Explain how you found your answer.

C. Formulas

Stacking Chairs

The picture below shows a stack of chairs. Notice that the height of one chair is 80 centimeters, and a stack of two chairs is 87 centimeters high.

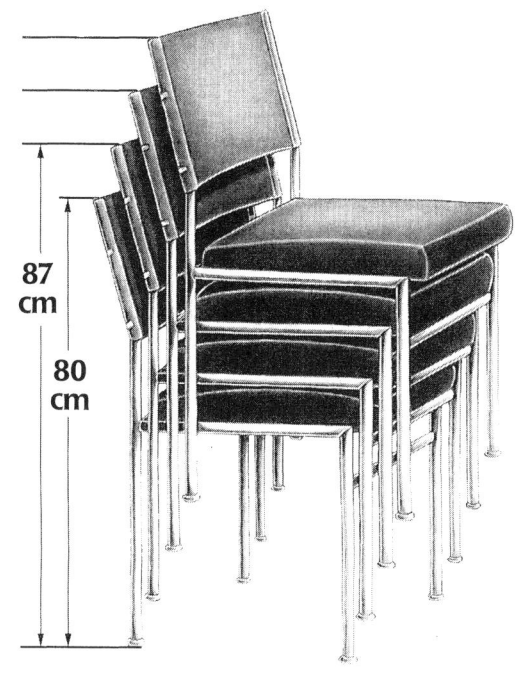

Damian suggests that the following formula can be used to find the height of a stack of these chairs:

number of chairs $\xrightarrow{-1}$ ___ $\xrightarrow{\times 7}$ ___ $\xrightarrow{+80}$ height

22. Explain what each of the numbers in the formula represents.

23. Alba thinks that a formula like this would do just as well:

number of chairs $\xrightarrow{\times}$ ___ $\xrightarrow{+}$ height

   a. What numbers would Alba use in her formula? Explain how you determined these numbers.

   b. Alba thought about making a formula like this:

   number of chairs $\xrightarrow{+}$ ___ $\xrightarrow{\times}$ height

   Will this work? Why or why not?

24. These chairs are used in an auditorium and sometimes have to be stored underneath the stage. The storage space is 116 centimeters high.

   a. How many chairs can be put in a stack that will fit in the storage space?

   b. Describe your calculation with an arrow string.

25. For another style of chair, there is a different formula:

   number of chairs $\xrightarrow{\times 11}$ ___ $\xrightarrow{+54}$ height

   a. How are these chairs different from the first ones?

   b. If the storage space were 150 centimeters high, would the following arrow string give the number of chairs that would fit? Why or why not?

   150 $\xrightarrow{\div 11}$ ___ $\xrightarrow{-54}$ ___

Mathematics in Context • Expressions and Formulas   21

Expressions and Formulas

Bike Sizes

You have discovered some formulas yourself. On the next two pages, you will use formulas that other people have developed.

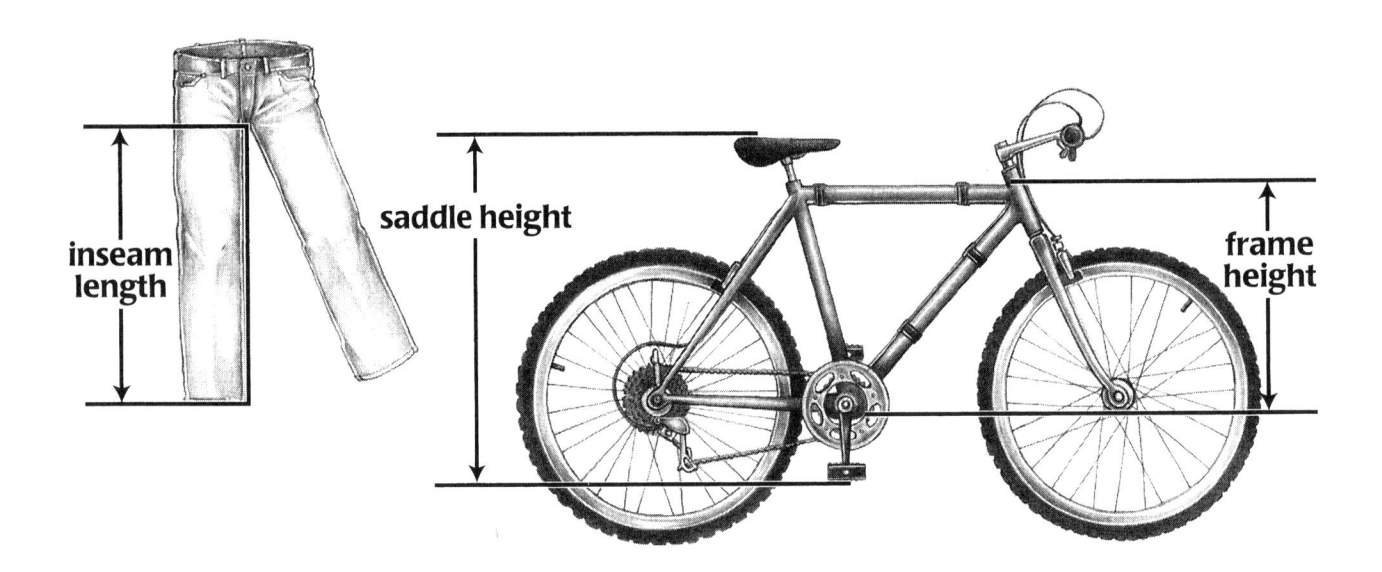

Stores that sell bicycles use formulas to find the best saddle and frame heights for each customer. One number used in these formulas is the inseam of the cyclist. This is the length of the cyclist's leg, measured in centimeters along the inside seam of the pants. The saddle height is calculated with the following formula:

> inseam (in cm) × 1.08 = saddle height (in cm)

26. Write an arrow string for this formula.

The formula for the frame height is as follows:

> inseam (in cm) × 0.66 + 2 cm = frame height (in cm)

27. Write an arrow string for this formula.

C. Formulas

Formulas are often written with the result first, for example:

> saddle height (in cm) = inseam (in cm) × 1.08
>
> frame height (in cm) = inseam (in cm) × 0.66 + 2 cm

28. Are the arrow formulas you wrote (for problems **26** and **27**) correct for these rewritten formulas above?

29. Carlie has an inseam of 70 centimeters. What frame and saddle heights does she need?

30. Look at the bike pictured below.

  a. What is the frame height?

  b. What is the saddle height?

  c. Do both of these numbers correspond to the same inseam length? How did you find your answer?

81 cm

54 cm

31. If you were buying a bike, what frame and saddle heights would you need?

Mathematics in Context • Expressions and Formulas

Expressions and Formulas

Are They the Same?

32. Below are two pairs of arrow strings. They look similar, but are they really the same? Try some different numbers for the input. Then explain why you think the arrow strings in each case are the same or different.

  a. input $\xrightarrow{+7}$ ___ $\xrightarrow{-2}$ output

  input $\xrightarrow{-2}$ ___ $\xrightarrow{+7}$ output

  b. input $\xrightarrow{\times 2}$ ___ $\xrightarrow{\div 10}$ output

  input $\xrightarrow{\div 10}$ ___ $\xrightarrow{\times 2}$ output

33. Compare the two strings below and decide whether they are the same or different.

  input $\xrightarrow{\times 4}$ ___ $\xrightarrow{+3}$ output

  input $\xrightarrow{+3}$ ___ $\xrightarrow{\times 4}$ output

34. Write about what you discovered while working on problems **32** and **33**. Think of a rule that explains what you found.

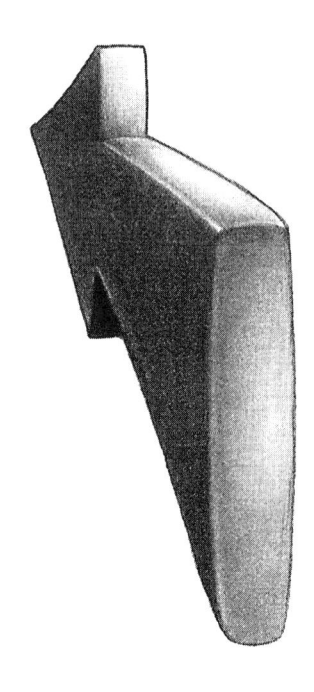

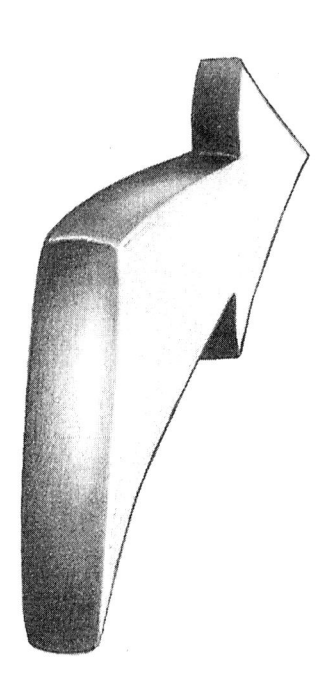

C. Formulas

Summary

A formula shows a procedure that can be used over and over again for different numbers in the same situation.

One formula used for bikes is:

inseam (in cm) × 0.66 + 2 cm = frame height (in cm)

or, written with the result first,

frame height (in cm) = inseam (in cm) × 0.66 + 2 cm

Many formulas can be described with arrow strings, for example:

inseam $\xrightarrow{\times 0.66}$ ___ $\xrightarrow{+2}$ frame height

Sometimes it is possible to change the order of the arrows in a string. If a problem has only addition and subtraction or only multiplication and division, the order can be changed, and the result will stay the same.

Summary Questions

35. Why is it useful to write a formula as an arrow string?

Expressions and Formulas

Summary Questions, continued

Mr. Macker is a band teacher who travels to different schools. Every day, he must decide how long it will take him to get to school and prepare to teach. He knows that he must allow 2 minutes for every mile between his home and a school. He also allows 15 minutes to set up the room before students arrive.

36. How much time should Mr. Macker allow if a school is 12 miles from his home?

37. Write a formula that Mr. Macker could use to find the time needed each day.

38. One day Mr. Macker decides that he needs 73 minutes. How far does he live from that school? Explain how you found your answer.

39. Which of the arrow strings labeled **a** and **b** will give the same result as the arrow string below? Explain.

$$\text{input} \xrightarrow{+2} 5 \xrightarrow{+3} 8 \xrightarrow{\times 4} \text{output}$$

a. $\text{input} \xrightarrow{+2} 5 \xrightarrow{\times 4} 20 \xrightarrow{+3} \text{output}$

b. $\text{input} \xrightarrow{+3} \underline{} \xrightarrow{+2} \underline{} \xrightarrow{\times 4} \text{output}$

D. REVERSE OPERATIONS

FOREIGN MONEY

Marty is going to visit The Netherlands and has to buy Dutch guilders. The exchange rate is 1.65 Dutch guilders for every one United States dollar.

This arrow string shows the calculation:

number of dollars $\xrightarrow{\times 1.65}$ number of guilders

1. How many Dutch guilders would Marty get for 50 United States dollars?

2. Copy and complete this table. Then explain how you found the numbers.

United States Dollars	1	2	3	4	5	6	7	8	9	10
Dutch Guilders										

3. In The Netherlands, Marty pays 2.45 Dutch guilders for a hamburger. Use the table to estimate about how much that would be in United States dollars.

4. **a.** Look closely at your table. Find a column in which the number of Dutch guilders is close to a whole number.

   b. How could this discovery be useful?

Mathematics in Context • Expressions and Formulas

Expressions and Formulas

To estimate prices in dollars so that he has an idea of how much he is paying, Marty uses the fact that 10 guilders is about 6 dollars. He thinks of the following rule:

number of guilders $\xrightarrow{\div 10}$ ___ $\xrightarrow{\times 6}$ number of dollars

5. Will this rule work? Explain your answer.

6. The prices for the items pictured below are given in Dutch guilders. Use the rule above to find the approximate cost of each item in United States dollars.

Nanda, Marty's Dutch friend, is coming to the United States for a visit. She wants to change Marty's conversion rule so that she can use it to find out how much things cost in guilders when the price is in dollars.

D. Reverse Operations

Marty's rule:

number of guilders $\xrightarrow{\div 10}$ ___ $\xrightarrow{\times 6}$ number of dollars

7. a. How can Marty's rule be changed to convert United States dollars to Dutch guilders? Write the new rule as an arrow string.

   b. Does your rule from part **a** agree with the exchange rate given at the beginning of this section?

Marty is surprised that the new rule seems so much harder than the one he used.

8. Why is the rule for converting United States dollars to Dutch guilders harder?

Nanda thinks of a simpler rule:

number of dollars $\xrightarrow{\times 3}$ ___ $\xrightarrow{\div 2}$ number of guilders

She also thinks of another rule:

number of dollars $\xrightarrow{\times 5}$ ___ $\xrightarrow{\div 3}$ number of guilders

9. How well do these rules work?

10. If you were Nanda, which rule would you use? Why?

11. Think of three things that someone visiting your city might want to buy. For each item, estimate the price in United States dollars. Then calculate how much it would cost in Dutch guilders. Use the exchange rate given in this section.

Lisa and Andre are pen pals. In a letter to Andre, Lisa wrote that she paid $5 to see a movie in the United States. Andre had seen the same movie in The Netherlands and paid 13 Dutch guilders.

12. In which country did the movie cost more? Show how you found your answer.

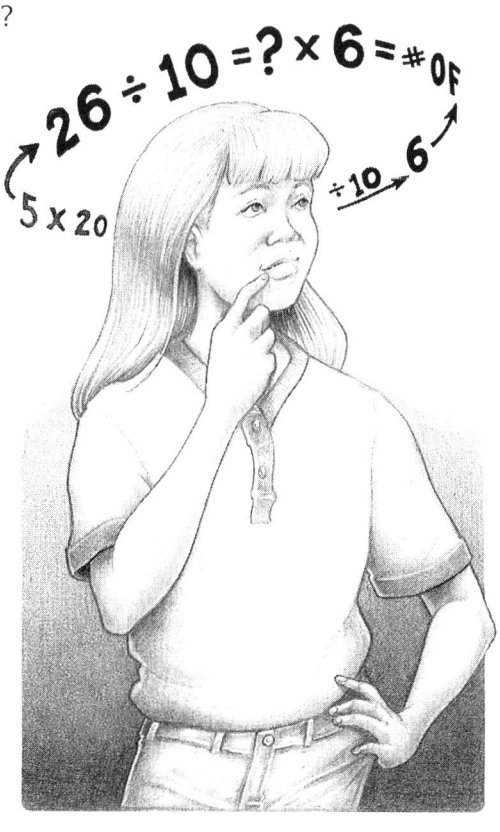

Mathematics in Context • Expressions and Formulas

Expressions and Formulas

GOING BACKWARDS

Pat and Kris are playing a game. One player writes down an arrow string and the output (answer) but not the input (starting number). The other player has to determine the input.

Here is Pat's arrow string and output:

$$\underline{\;?\;} \xrightarrow{+4} \underline{\quad} \xrightarrow{\times 10} \underline{\quad} \xrightarrow{-2} \underline{\quad} \xrightarrow{\div 2} 29$$

13. a. What should Kris give as the input? Explain how you came up with this number.

b. One student found an answer for Kris by using a *reverse string*. What should go above each of the reversed arrows below?

$$\underline{\overset{5}{?}} \xleftarrow{\div 5} \underline{20} \xleftarrow{+8} \underline{28} \xleftarrow{-5} \underline{23} \xleftarrow{+6} 29$$

On the next round, Kris wrote:

$$\underline{\overset{15}{?}} \xrightarrow{+3} \underline{18} \xrightarrow{\div 6} \underline{3} \xrightarrow{+5} \underline{8} \xrightarrow{+2} 6$$

14. a. What should Pat give as the input? Explain how you found this number.

b. Write the reverse string that can be used to find the input.

D. Reverse Operations

Beech Trees

There are some beech trees in a park near Jessica's house. Over the years, a number of botanists have studied these trees very carefully. They discovered something very interesting. When a tree is between 20 and 80 years old, it grows fairly evenly. They developed two formulas that describe the growth of the trees if the age is known.

age $\xrightarrow{\times 0.4}$ ___ $\xrightarrow{-2.5}$ thickness   age $\xrightarrow{\times 0.4}$ ___ $\xrightarrow{+1}$ height

For these formulas, the age is in years; the height is in meters; the thickness is in centimeters and is measured 1 meter from the ground.

15. Find the heights and thicknesses of trees that are 20, 30, and 40 years old.

16. Jessica is interested in knowing the age of a tree. How can she find it?

Thickness refers to the diameter of the tree trunk. Jessica uses some straight sticks to help her measure the thickness of a tree. She finds that it is 25.5 centimeters.

17. How old is the tree?

Jessica estimates the height of another beech tree to be about 20 meters.

18. Use this estimate of the height to calculate the age of the tree.

Jessica realizes that she can make a new formula. Her new formula gives the height of a tree if the thickness is known.

19. Write Jessica's formula.

Mathematics in Context • Expressions and Formulas

Expressions and Formulas

Ham and Cheese

Carmen and Andy are at the store buying ham and cheese for sandwiches. Carmen is going to buy Swiss cheese that costs $4.40 per pound. She decides that she wants to buy 0.75 pound. She wants to know how much it will cost before she orders it.

20. Write an arrow string to show how much it will cost.

This is the string that Carmen wrote:

$$\$4.40 \xrightarrow{\div 4} \$1.10 \xrightarrow{\times 3} \$3.30$$

21. Is Carmen's string correct? What did she do?

Carmen meets Andy, who has already bought some ham. Andy bought 0.75 pound, and it cost $2.25. Carmen wonders what the price per pound is.

22. Write an arrow string to find the price per pound of ham.

Summary

Every arrow has a reverse arrow. A reverse arrow has the opposite operation.

For example, the reverse of $\xrightarrow{\div 4}$ is $\xleftarrow{\times 4}$.

Reverse arrows can be used to make reverse strings. For example,

___ $\xrightarrow{\div 4}$ ___ $\xrightarrow{\times 3}$ ___ reverses to ___ $\xleftarrow{\times 4}$ ___ $\xleftarrow{\div 3}$ ___ ,

which is the same as ___ $\xrightarrow{\div 3}$ ___ $\xrightarrow{\times 4}$ ___ .

Summary Questions

23. When are reverse strings useful?

24. Write the reverse for each of the following strings:

  a. input $\xrightarrow{+2}$ ___ $\xrightarrow{\times 3}$ ___ $\xrightarrow{-4}$ output

  b. input $\xrightarrow{\div 2}$ ___ $\xrightarrow{-5}$ ___ $\xrightarrow{+7}$ output

  c. Use some numbers to find out whether or not your reverse strings are correct.

E. TABLES

Apple Crisp

APPLE CRISP—*serves four*

You will need:

- 250 grams of flour
- 125 grams of butter
- 1 egg
- 450 grams of apples
- water, salt, cinnamon, lemon juice, and sugar

The chef at the Very Good Restaurant bakes apple crisp using a recipe in the Very Good Cookbook. The size of the apple crisp is determined by the number of people it must serve.

The chef has to calculate the amounts of ingredients needed for different sizes of apple crisp. The original recipe makes an apple crisp that will serve four people. One day, the chef receives an order for an apple crisp to serve six people. The chef begins the calculations by finding the amounts of ingredients needed to serve two people.

Ingredients	4 servings	2 servings	6 servings
Flour	250 g		
Butter	125 g		
Eggs	1		
Apples	450 g		

Use **Student Activity Sheet 1** to answer problems **1–3**.

1. **a.** Fill in the missing values in the first table on **Student Activity Sheet 1**.
   b. Why do you think that the chef started with a two-serving apple crisp?

Ingredients	Servings								
	2	3	4	5	6	8	10	12
Flour			250 g					
Butter			125 g					
Eggs			1					
Apples			450 g					

2. Fill in the second table on **Student Activity Sheet 1**.

3. Describe some of the strategies you used to fill in the table.

Mathematics in Context • Expressions and Formulas

Expressions and Formulas

Shuttle Service

The Lake Shore Bus Company runs a shuttle service between Chicago and northwest Indiana. Steve has a copy of the timetable for the service.

The bus company decides to make a new stop at the Expo Center.

O'HARE TO NORTHWEST INDIANA

Find the time you leave O'Hare in the left column.
Read straight across on the same line to your destination point.
This will show your arrival time at that point.

Leave O'Hare Lower Terminal	Arrive Expo Center	Arrive Hammond/ Highland	Arrive Glen Park	Arrive Merrillville
5:45 A.M.		7:05 A.M.	7:20 A.M.	7:40 A.M.
6:45 A.M.		8:05 A.M.	8:20 A.M.	8:40 A.M.
7:45 A.M.		9:05 A.M.	9:20 A.M.	9:40 A.M.
8:45 A.M.		10:05 A.M.	10:20 A.M.	10:40 A.M.
9:45 A.M.		11:05 A.M.	11:20 A.M.	11:40 A.M.
10:45 A.M.		12:05 P.M.	12:20 P.M.	12:40 P.M.
11:45 A.M.		1:05 P.M.	1:20 P.M.	1:40 P.M.
12:45 P.M.		2:05 P.M.	2:20 P.M.	2:40 P.M.
1:45 P.M.		3:05 P.M.	3:20 P.M.	3:40 P.M.
2:45 P.M.		4:05 P.M.	4:20 P.M.	4:40 P.M.

Steve's job is to add the new column for the Expo Center stop. Luckily, he knows that the bus arrives at the Expo Center 15 minutes after it leaves O'Hare's Lower Terminal.

4. Use the timetable on **Student Activity Sheet 2** to fill in the times for the Expo Center stop.

E. Tables

You may have noticed that the second column can be filled in by adding 15 minutes to the times in the first column. This can be shown with an arrow string.

$$\text{Lower Terminal} \xrightarrow{+ 15 \text{ minutes}} \text{Expo Center}$$

5. The other columns can be created in much the same way. Copy and complete the string below.

$$\text{Lower Terminal} \xrightarrow{+ 15 \text{ minutes}} \text{Expo Center} \xrightarrow{?} \text{Hammond} \xrightarrow{?} \text{Glen Park} \xrightarrow{?} \text{Merrillville}$$

6. a. There is also a pattern going down each column. How can this regularity be explained?

b. Use the pattern to add an extra row to the timetable on **Student Activity Sheet 2.**

The timetable for the Lake Shore Bus Company can be generated with two arrow strings: a horizontal arrow string and a vertical arrow string.

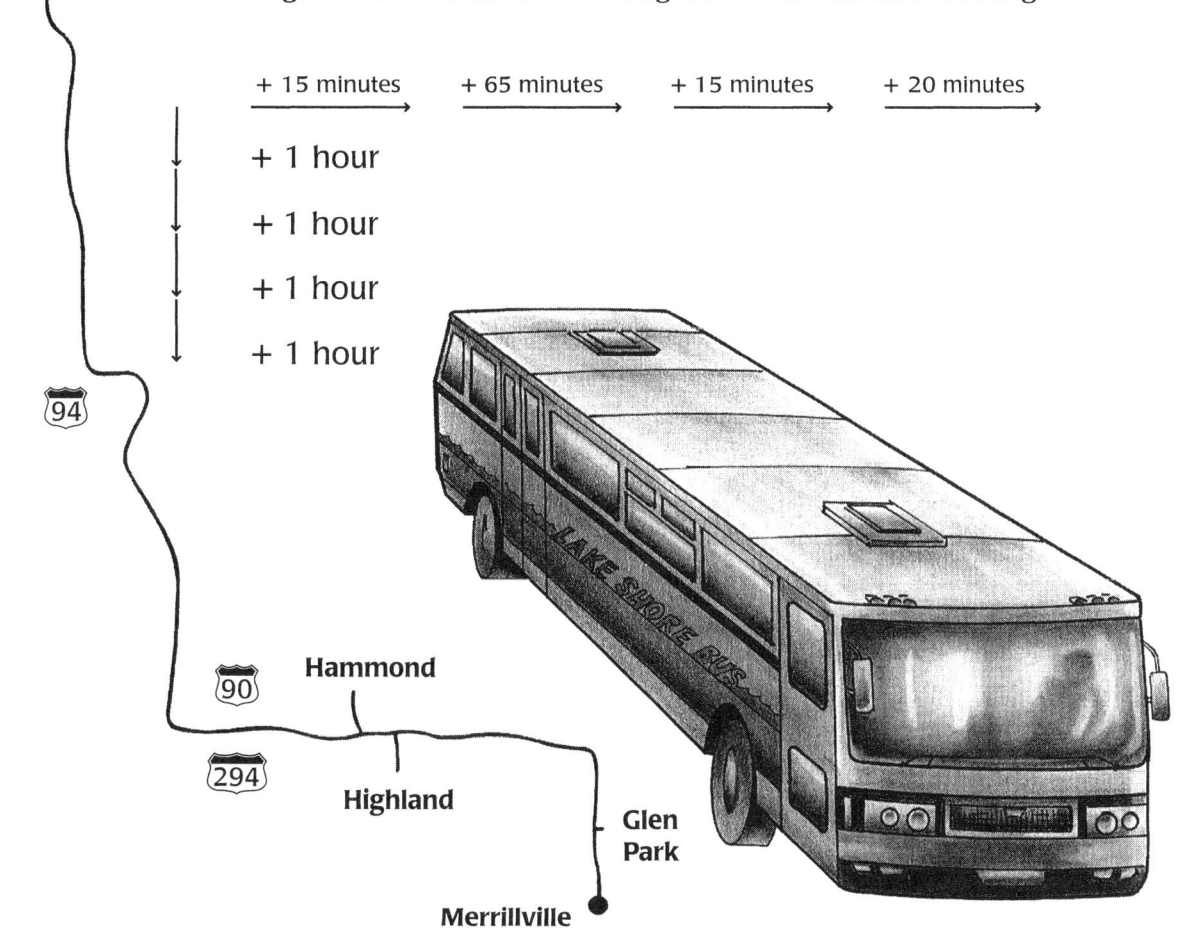

$$\xrightarrow{+ 15 \text{ minutes}} \quad \xrightarrow{+ 65 \text{ minutes}} \quad \xrightarrow{+ 15 \text{ minutes}} \quad \xrightarrow{+ 20 \text{ minutes}}$$

+ 1 hour
+ 1 hour
+ 1 hour
+ 1 hour

Mathematics in Context • Expressions and Formulas

Expressions and Formulas

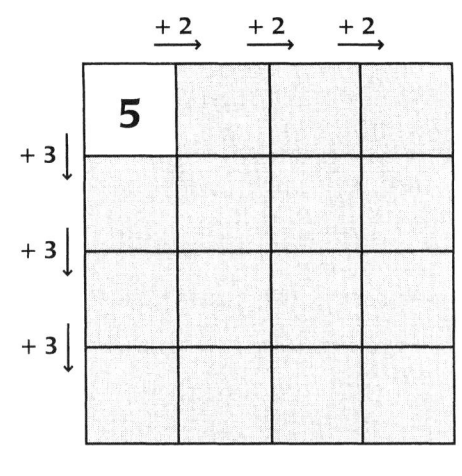

Like the timetable for the bus, other tables can be created from a pair of strings.

7. Complete the table at the top of **Student Activity Sheet 3**.

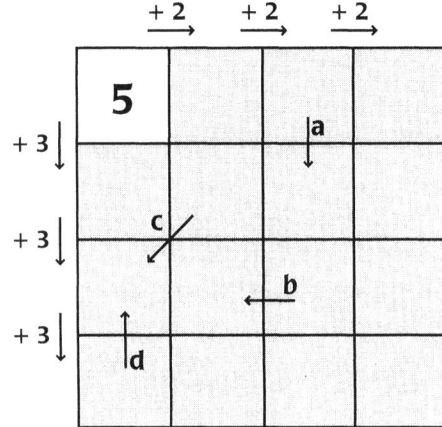

8. What operations fit with the arrows marked **a** through **d** on the table in the middle of **Student Activity Sheet 3**?

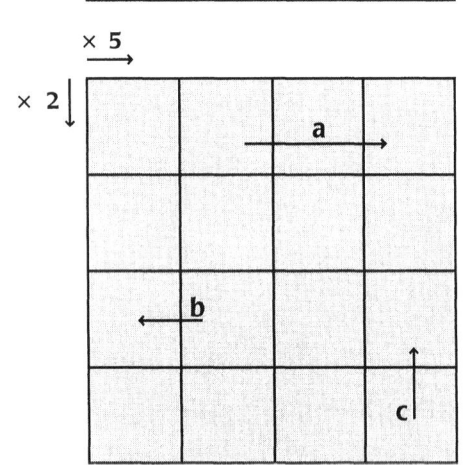

Instead of having an arrow for each row and column, a table can be written with one arrow to show the change for every row, and one arrow to show the change for every column.

The table on the left can be filled in by multiplying by five for each move to the right and by two for each move down.

9. Which operations fit with the arrows marked **a** through **c** inside the table?

× 5 →

× 2 ↓

	700	

10. Only one number in the table on the left is known. Write the missing numbers in the table at the bottom of **Student Activity Sheet 3.**

+ 4 →

× 3 ↓

2		

The table on the left is different from the previous ones.

11. Copy and complete the table.

12. Compare your answers with those of your classmates. Do some students have different numbers in the table? If so, why?

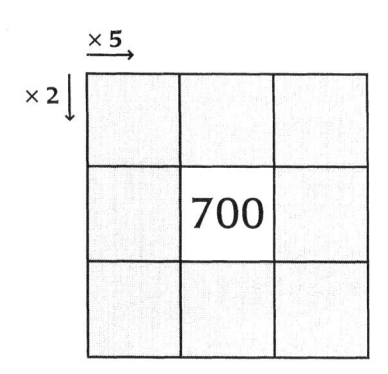

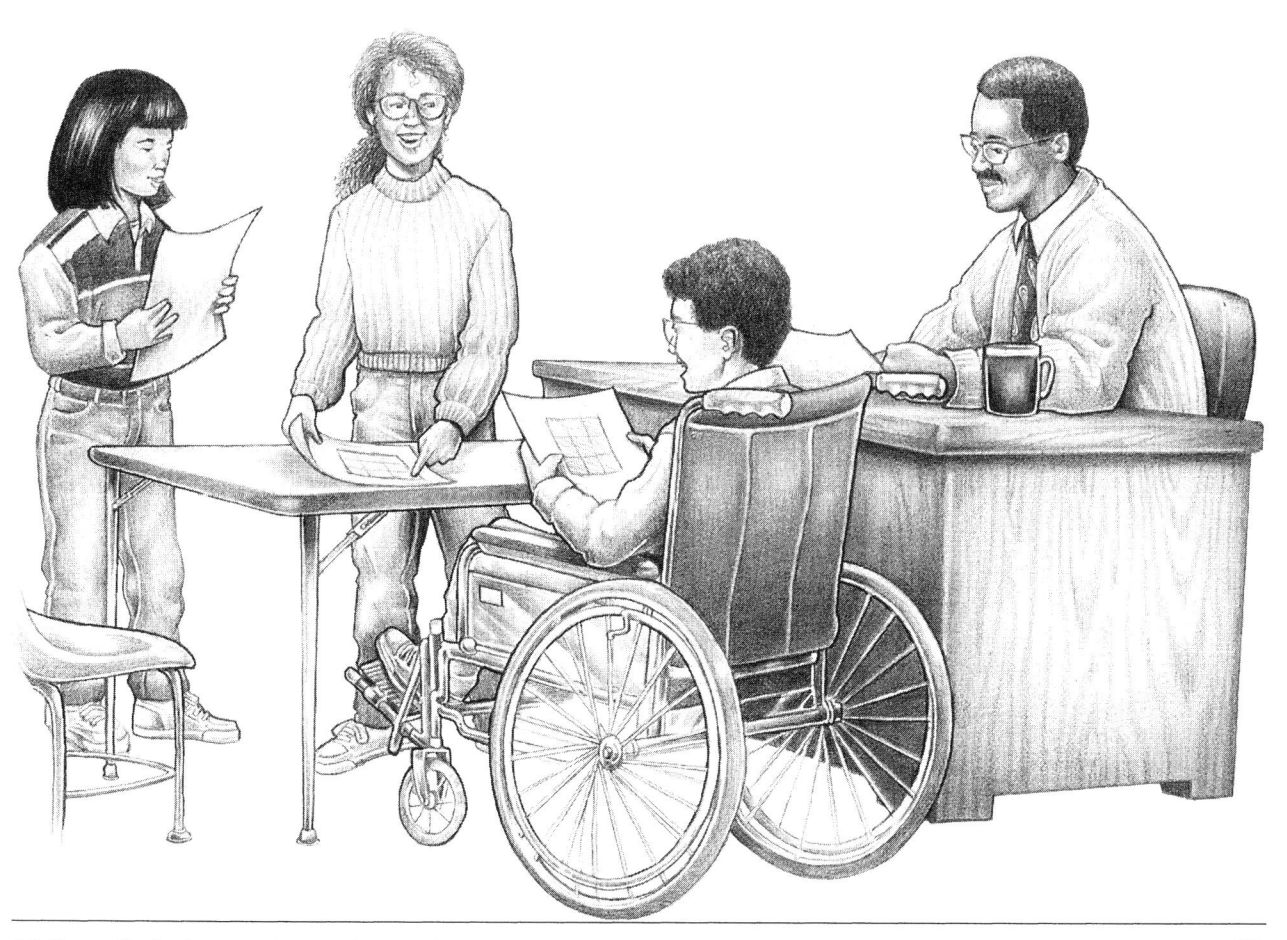

Expressions and Formulas

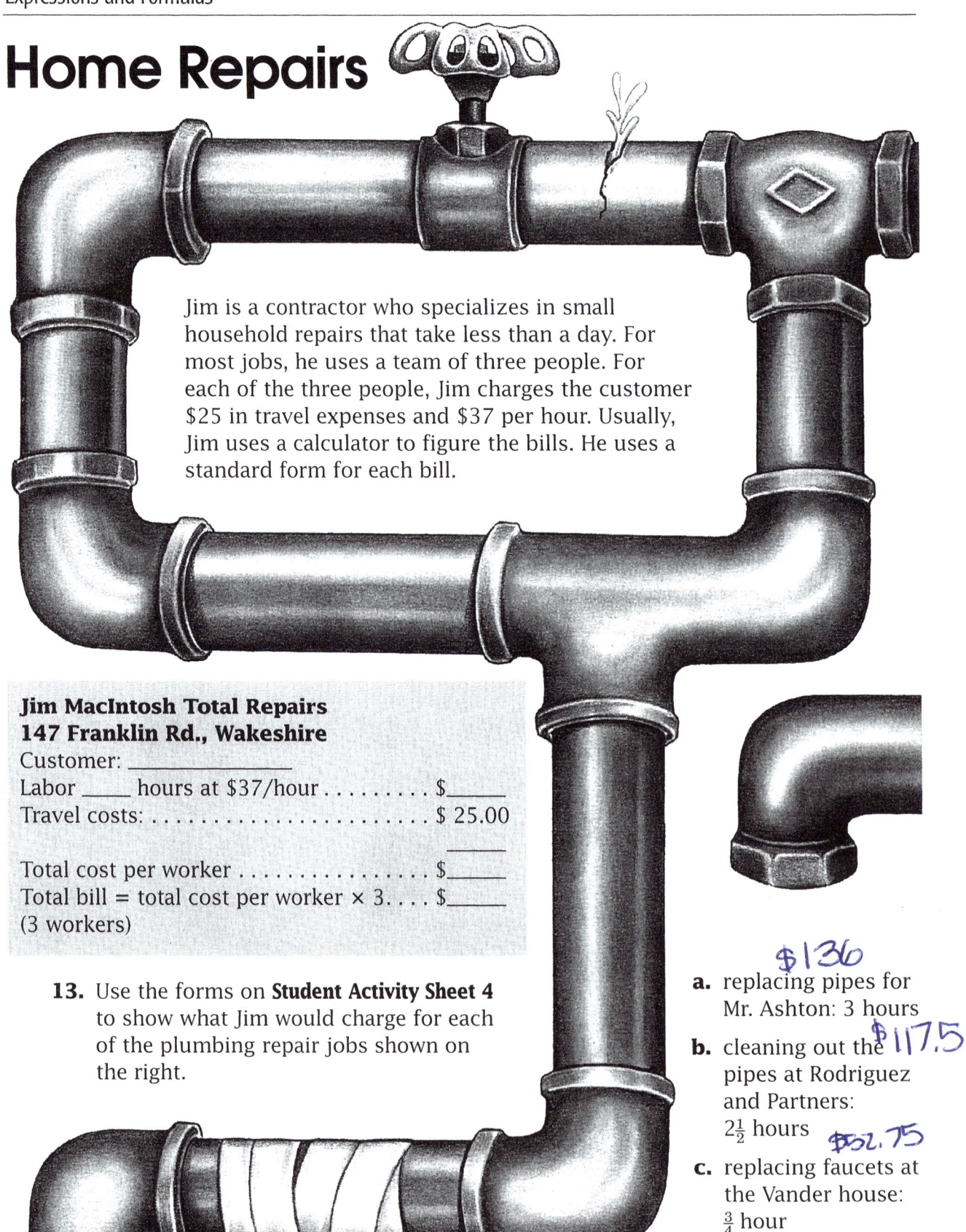

Home Repairs

Jim is a contractor who specializes in small household repairs that take less than a day. For most jobs, he uses a team of three people. For each of the three people, Jim charges the customer $25 in travel expenses and $37 per hour. Usually, Jim uses a calculator to figure the bills. He uses a standard form for each bill.

Jim MacIntosh Total Repairs
147 Franklin Rd., Wakeshire
Customer: _____
Labor ____ hours at $37/hour .... $ _____
Travel costs: .... $ 25.00
                                  _____
Total cost per worker .... $ _____
Total bill = total cost per worker × 3 .... $ _____
(3 workers)

13. Use the forms on **Student Activity Sheet 4** to show what Jim would charge for each of the plumbing repair jobs shown on the right.

a. replacing pipes for Mr. Ashton: 3 hours $136

b. cleaning out the pipes at Rodriguez and Partners: $117.5
$2\frac{1}{2}$ hours

c. replacing faucets at the Vander house: $52.75
$\frac{3}{4}$ hour

E. Tables

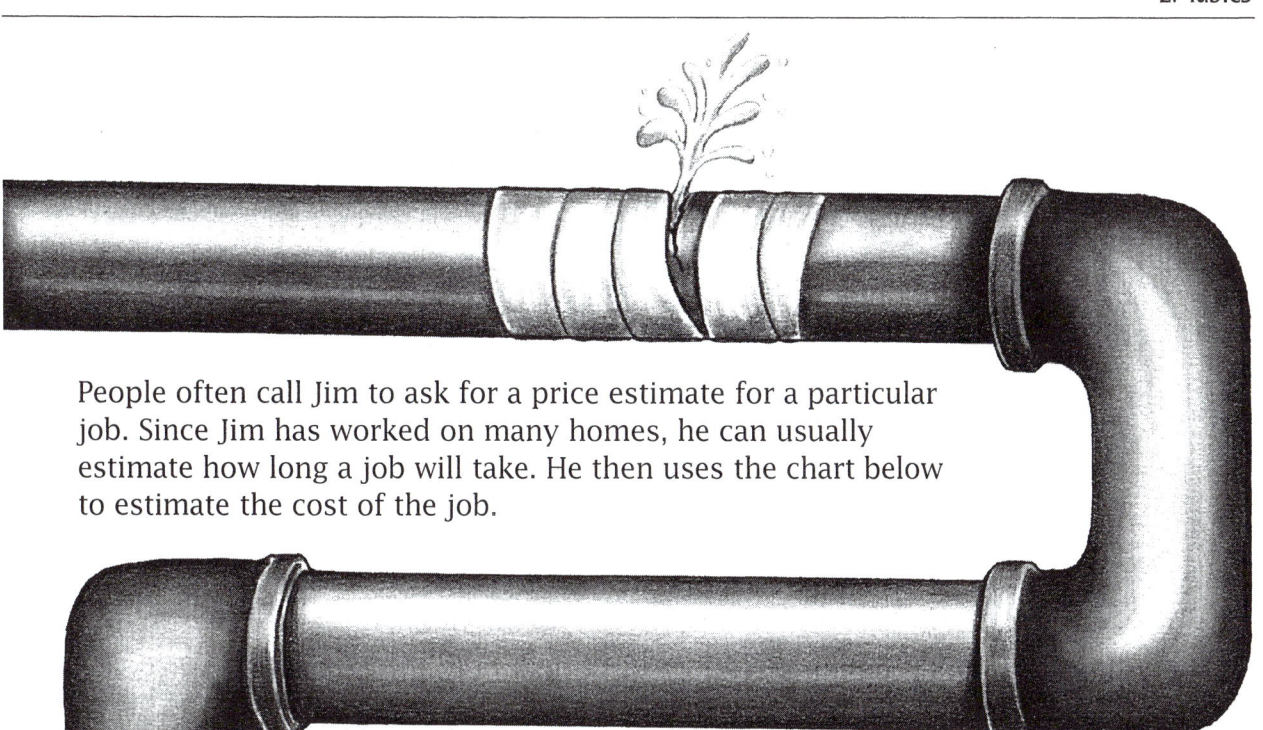

People often call Jim to ask for a price estimate for a particular job. Since Jim has worked on many homes, he can usually estimate how long a job will take. He then uses the chart below to estimate the cost of the job.

Hours	Labor Cost per Worker (in dollars)	Travel Cost per Worker (in dollars)	Cost per Worker (in dollars)	Total for Three Workers (in dollars)
1	37	25	62	186
2	74	25	99	297
3	111	25	136	408
4	148	25	173	519

14. a. What do the entries in the first row of the table mean? *show the estimate for 1 hour of work*

   b. What would the row for 5 hours look like? *5, 185, 25, 210, 630*

15. a. Explain the regularity in the column for labor cost per worker. *It will always be that because the travel cost is the same*

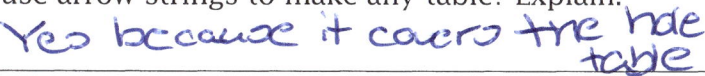

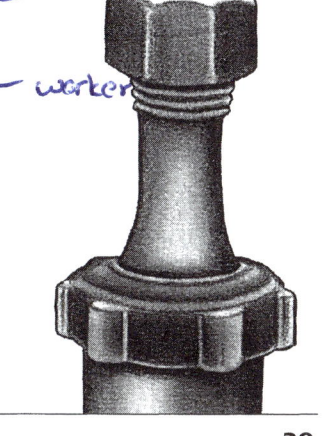

   every time they add 7 × hours to get the labor

   b. Look carefully at the table. Make a list of all of the regularities you can find. Explain the regularities.

   Hours ─×37→ Labor Cost ─+25→ Cost per worker
   ─×3→ Total for 3 workers

16. a. Draw an arrow string that Jim could use to make more rows for the table.

   b. Use your arrow string to make two more rows (for 6 and 7 hours) for the table.

17. Can you use arrow strings to make any table? Explain.
   Yes because it covers the hole table

Mathematics in Context • Expressions and Formulas

Expressions and Formulas

Look carefully at the following table.

Typical Average Temperature (°F)												
	J	F	M	A	M	J	J	A	S	O	N	D
Acapulco	88	88	88	88	89	90	91	91	90	90	90	89
Antigua	80	80	80	82	90	90	90	90	89	89	89	83
Aruba	83	84	84	86	88	88	88	91	91	90	89	86
Cancún	84	85	88	91	94	92	92	91	90	88	86	82
Cozumel	84	85	88	91	94	92	92	91	90	88	86	82
Grand Cayman	88	87	86	88	88	89	90	91	91	89	88	88
Ixtapa	89	90	92	93	89	88	89	90	91	91	90	89
Jamaica	86	87	87	88	90	90	90	90	89	89	89	87
Los Cabos	73	74	79	83	88	93	95	93	92	89	82	74
Manzanillo	77	78	82	86	83	88	93	95	93	92	89	74
Mazatlán	73	74	79	83	84	92	94	92	92	90	85	71
Nassau	76	76	78	80	84	88	89	90	88	84	81	79
Puerto Vallarta	76	77	81	85	83	88	93	95	93	92	89	75
St. Martin/St. Kitts	80	81	82	83	86	86	86	87	87	86	85	84
U. S. Virgin Islands	80	81	82	83	88	88	90	90	88	87	86	86

18. a. Try to find the numbers and operations used to make this table. Explain what you found and if it makes sense.

b. Do you need to change your answer to problem **17**?
No   Yes

Summary

Some tables have regularities that can be described by a horizontal arrow string and a vertical arrow string. The strings can be useful in making or extending a table.

Be careful! Not every table has regularities.

Summary Questions

19. Find two tables from a newspaper or magazine, one table that has regularities and one that does not.

20. How can you tell whether arrow strings can be used to describe a table? *By if you see a pattern in the table*

F. ORDER OF OPERATIONS

Arithmetic Trees

While working on a problem from the previous section, a student named Enrique wrote the arrow string below. The problem was to find the cost of having three workers for two hours of repairs.

$$2 \xrightarrow{\times 37} \underline{} \xrightarrow{+ 25} \underline{} \xrightarrow{\times 3} \underline{}$$

Karlene was working with Enrique, and she wrote:

$$2 \times 37 + 25 \times 3$$

She solved the problem and got an answer of 149. Her friend Enrique was very surprised.

1. a. How did Karlene get 149?

   b. Why was Enrique surprised?

Karlene and Enrique decided that the number sentence 2 × 37 + 25 × 3 is not necessarily the same as the arrow string:

$$2 \xrightarrow{\times 37} \underline{} \xrightarrow{+ 25} \underline{} \xrightarrow{\times 3} \underline{}$$

There is more than one way to interpret the number sentence. The calculations can be done in different orders.

2. Solve each of the problems below and compare your answers to those found by other students in your class.

   a. 1 + 11 × 11

   b. 10 × 10 + 1

   c. $10 \xrightarrow{\times 10} \underline{} \xrightarrow{+ 2} \underline{}$

   d. When can you be sure that everyone will get the same answer?

Sometimes the context of a problem helps you understand how to calculate it. For instance, in the repairs problem, Karlene and Enrique knew that the 3 represented the number of workers. So it makes sense to first calculate the subtotal of 2 × 37 + 25 and then multiply the result by three.

Sometimes people write their calculations for a problem in a very poor way:

$$2 \times 37 = 74 + 25 = 99 \times 3 = 297$$

3. Why is this a poor way to write the calculation?

Mathematics in Context • Expressions and Formulas

Expressions and Formulas

So that everyone can come up with the same answer to problems with different operations in them, mathematicians decided to treat multiplication and division as stronger than addition or subtraction. This means that you should do any multiplication or division before you add or subtract.

4. Use the mathematicians' way to find the value for each of the following expressions:

  a. 32 + 5 × 20

  b. 18 ÷ 3 + 2 × 5

  c. 47 − 11 + 6 × 8

Calculators and computers usually follow the mathematicians' rule. Some calculators, however, do not use the rule.

5. a. Use the mathematicians' rule to find 5 × 5 + 6 × 6 and 6 × 6 + 5 × 5.

  b. Does your calculator use the mathematicians' rule? How did you decide?

  c. Why do you think calculators have built-in rules?

It is important to have a way to write expressions so that it is clear which calculations to do first, second, and so on.

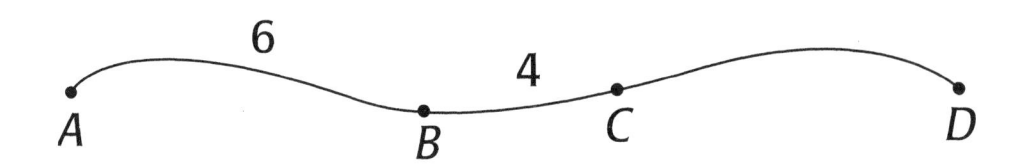

Above is a very simple map, not drawn to scale. Suppose someone told you that the distance from *A* to *D* is 15 miles. You can see in the drawing that the distance from *A* to *B* is 6 miles, and the distance from *B* to *C* is 4 miles.

6. What is the distance from *C* to *D*? Write down your calculations.

F. Order of Operations

Telly found the distance from *C* to *D* by adding 6 and 4. Then she subtracted the result from 15. She could have used an *arithmetic tree* to write down this calculation.

- To make an arithmetic tree, begin by writing all the numbers.

    15   6   4

- Then pick two numbers. Telly picked 6 and 4.

- Telly added the numbers and got 10.

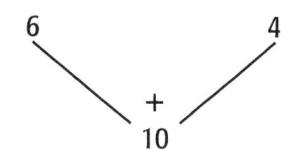

- Telly used the 15 and the new 10.

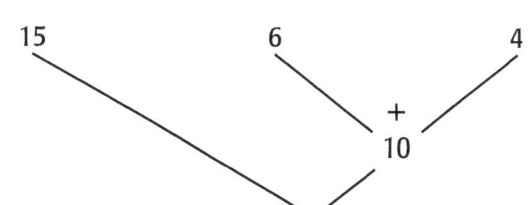

- She subtracted to get 5.

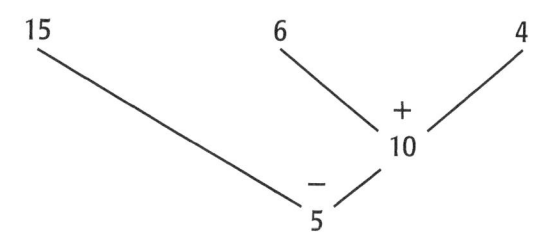

7. To help you understand arithmetic trees, complete those on **Student Activity Sheet 5**.

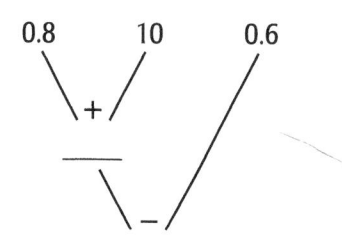

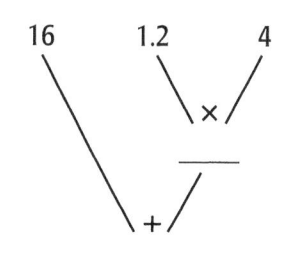

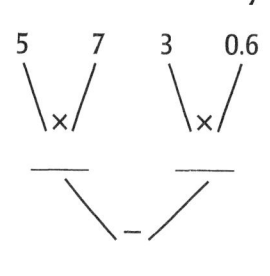

Mathematics in Context • Expressions and Formulas

Expressions and Formulas

8. Remember that multiplication is stronger than addition. Which of the following trees shows the proper calculation of 1 × 2 + 3 × 4?

   a.   **b.** 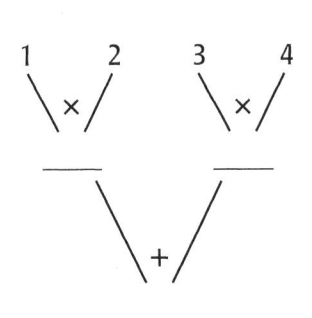   **c.**

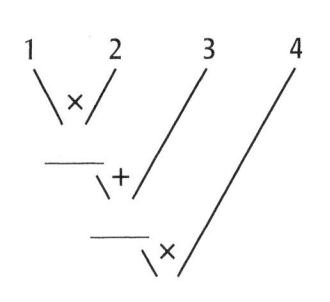

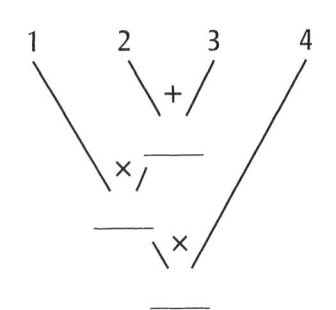

9. Which of the trees from problem **8** shows the same calculation as the arrow string below?

$$1 \xrightarrow{\times 2} 2 \xrightarrow{+3} 5 \xrightarrow{\times 4} 20$$

10. At the beginning of this section, two students—Karlene and Enrique—got different answers for the cost of a two-hour job.

   a. Does the tree below give the same answer that Karlene got?

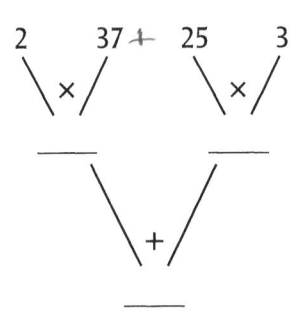

   b. Draw a tree that gives the correct answer for a two-hour job.

F. Order of Operations

One student feels that the calculation for repair bills (shown on pages 38 and 39) should start with the travel costs since the customer always has to pay them. Another student thinks that it is impossible to start an arrow string with the travel costs.

11. Is it impossible to start an arrow string with the travel costs? Explain.

An arithmetic tree can have the travel costs first.

12. Below is an arithmetic tree for calculating the bill for a two-hour job with the travel costs first. Does this arithmetic tree give the correct total (as shown in the table on page 39)?

```
    25    2   37    3
          \ × /
           74
       \ + /
         99
          \ × /
           297
```

Repair Bill $25.00

2 hrs × $37.00

Mathematics in Context • Expressions and Formulas 45

Expressions and Formulas

It is possible to build an arithmetic tree that uses words instead of numbers. In the left column of the table below, a word tree has been built to describe the repair bill. In the right column, numbers have been used.

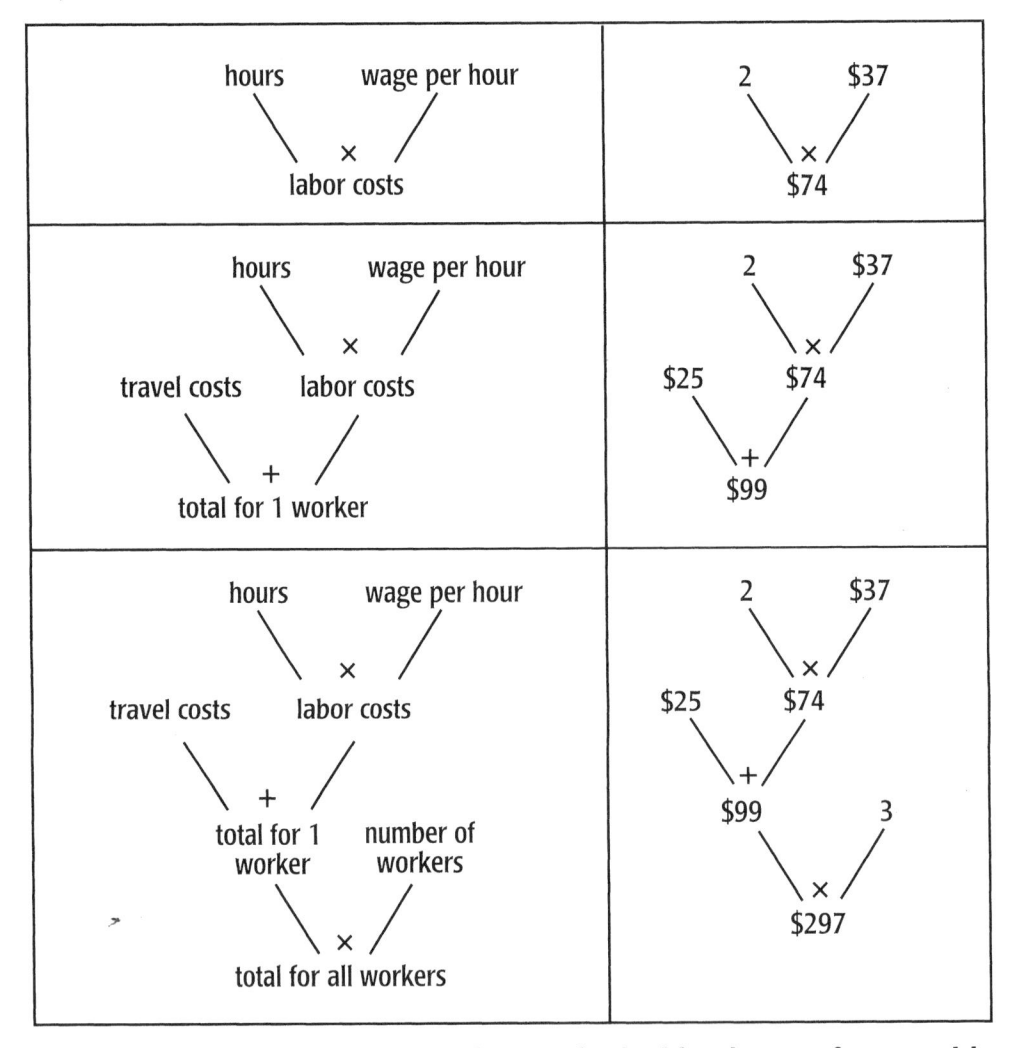

hours \ × / wage per hour → labor costs	2 \ × / $37 → $74
travel costs \ + / (hours × wage per hour → labor costs) → total for 1 worker	$25 \ + / (2 × $37 → $74) → $99
(travel costs + (hours × wage per hour → labor costs) → total for 1 worker) × number of workers → total for all workers	($25 + (2 × $37 → $74) → $99) × 3 → $297

By making some of the "branches" longer, the tree looks like the one from problem **12**.

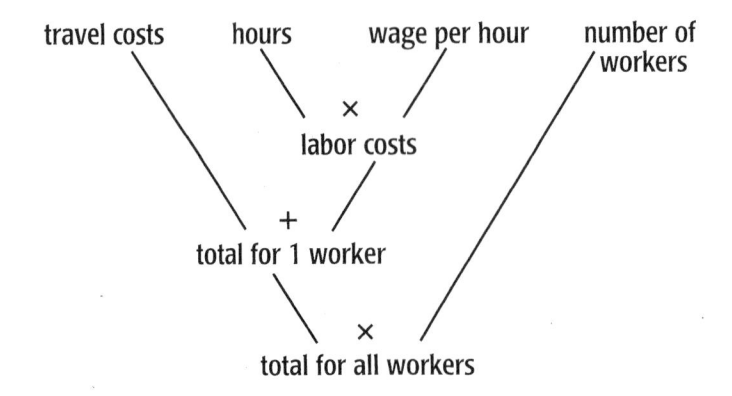

travel costs, hours, wage per hour, number of workers
 → labor costs (hours × wage per hour)
 → total for 1 worker (travel costs + labor costs)
 → total for all workers (total for 1 worker × number of workers)

13. Make a word tree that shows how to find the height of a stack of cups or chairs.

F. Order of Operations

Flexible Computation

Arithmetic trees can be used to help make some addition and subtraction problems easier.

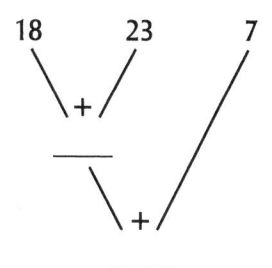

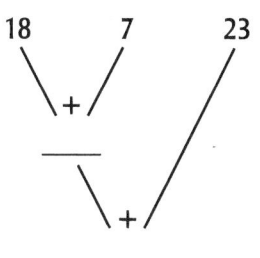

14. a. Compare the two trees above.

b. Draw a tree that makes adding the three numbers in part **a** easier.

Addition problems with more numbers have many possible arithmetic trees. Below are two trees for $\frac{1}{2} + \frac{1}{4} + \frac{3}{4} + \frac{3}{2}$.

15. a. Copy the trees and find the sum.

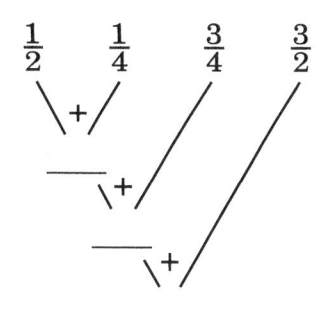

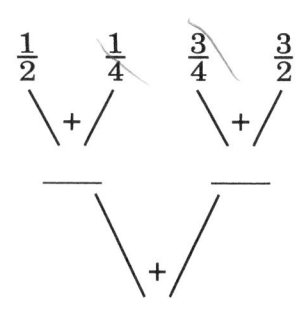

b. Design two other arithmetic trees for the same problem and solve them.

c. Which arithmetic tree makes $\frac{1}{2} + \frac{1}{4} + \frac{3}{4} + \frac{3}{2}$ easiest to compute? Why?

Expressions and Formulas

16. Design an arithmetic tree that will make each of the following problems easy to solve.

  a. 7 + 3 + 6 + 4

  b. 4.5 + 8.9 + 5.5 + 1.1

  c. $\frac{4}{10} + \frac{1}{2} + \frac{1}{10} + \frac{3}{4}$

17. How are different arithmetic trees for the same problem the same? How can they differ?

You may have noticed that if a problem has only addition, the answer is the same no matter how you make the arithmetic tree. You may wonder if this is true for subtraction.

18. Do the following trees give the same result?

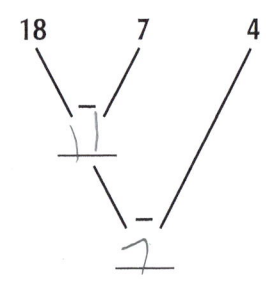

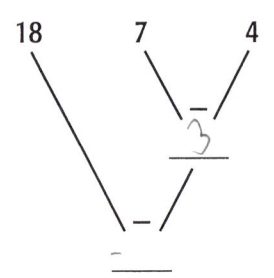

Near the beginning of the section, Telly used the following tree to find the distance from point *C* to point *D*:

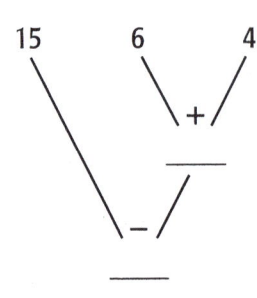

19. Make another arithmetic tree with 15, 6, and 4 across the top that gives the same result.

20. Calculate 176 − 89 − 11 and describe what you did. (*Hint:* If you look at the previous problem, you may think of an easy way to calculate this one.)

48   Britannica Mathematics System

F. Order of Operations

Return to the Supermarket

TOMATOES $1.50/lb

GRAPES $1.70/lb

GREEN BEANS $0.90/lb

The machine at Veggies-R-Us is broken. Ms. Prince buys 0.5 pound of grapes and 2 pounds of tomatoes.

21. How much does she have to pay?

22. Can you write an arrow string to show how to figure Ms. Prince's bill? Why or why not?

23. Can Ms. Prince's bill be calculated with an arithmetic tree? If so, make the tree. If not, explain why not.

Dr. Keppler buys 2 pounds of tomatoes, 0.5 pound of grapes, and $\frac{1}{2}$ pound of green beans.

24. Make an arithmetic tree for the total bill.

Mathematics in Context • Expressions and Formulas

Expressions and Formulas

25. Make a tree that can be used for any combination of tomatoes, grapes, and green beans. Use the words "weight of tomatoes," "weight of grapes," and so on.

The store manager gave each of the cashiers a calculator that uses the rule that multiplication is stronger than addition. Then she wrote these directions.

amount of tomatoes × 1.50 + amount of grapes × 1.70 + amount of green beans × 0.90 =
    (in pounds)   (in pounds)   (in pounds)

26. If the cashiers punch in a calculation using the directions above, will they get the correct total for the bill?

F. Order of Operations

WHAT COMES FIRST?

Arithmetic trees are useful because there is no question about the order of the calculation. The problem is that they take up a lot of room on paper. Copy the first tree on the right.

Since the 6 + 4 is done first, circle it on your copy.

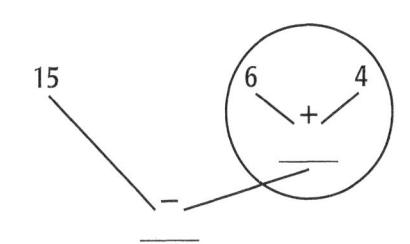

The tree can then be simplified to:

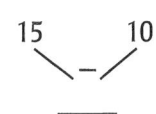

Instead of the second arithmetic tree, you could write: 15 − (6 + 4)

27. What does the circle mean?

The whole circle is not necessary. People often write 15 − (6 + 4). This does not take up much space, but it shows the order with the parentheses.

28. a. Rewrite the tree on the right using parentheses.

  b. Make a tree for 3 × (6 + 4).

  c. What is the value of 5 × (84 − 79)?

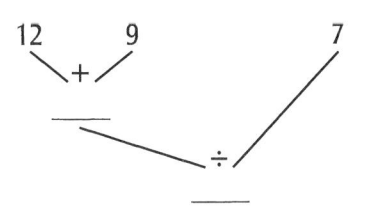

  d. Rewrite the tree on the lower right using parentheses.

29. Rewrite Karlene's problem from page 41: 2 × 37 + 25 × 3. Use parentheses so that the correct total for the bill will be calculated.

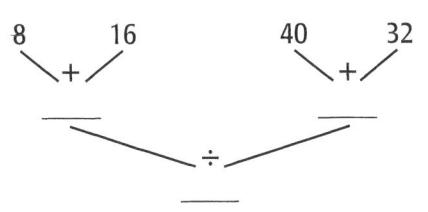

Mathematics in Context • Expressions and Formulas 51

Expressions and Formulas

Summary

The beginning of this unit focused on formulas using arrow language. There are also other ways to write formulas.

You can write them with words:

cost = tomatoes × $1.50 + grapes × $1.70 + green beans × $0.90
        (in lb)   (in lb)   (in lb)

You can use arithmetic trees:

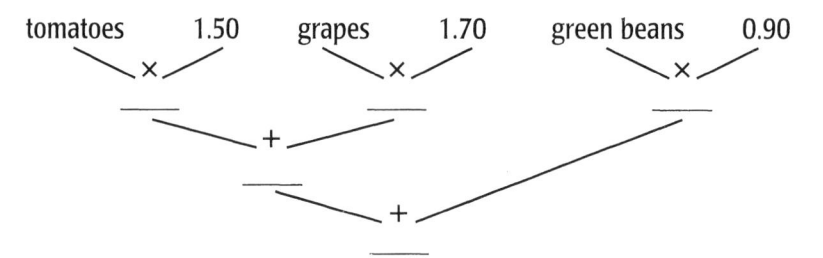

tomatoes   1.50   grapes   1.70   green beans   0.90
          ×   ×   ×
                +
                      +

Arithmetic trees show the order of calculation. If a problem is not in an arithmetic tree and does not have parentheses, there is a rule for the order of operations: Do the multiplication and division first.

$1 \times 2 + 3 \div 4 - 5$ is calculated as follows:

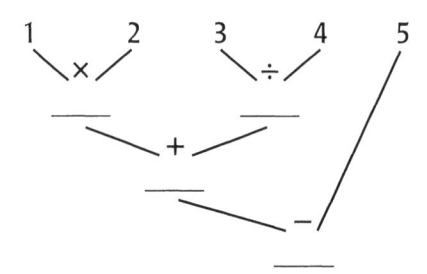

    1   2   3   4   5
     ×   ÷
        +
            −

Parentheses can be used to convert from an arithmetic tree to an expression that shows which operations to do first.

Summary Questions

30. Write a story in your journal in which the calculations below are used.

  a. (15 × 16 + 20) × 5

  b. 15 × 16 + 20 × 5

31. Phoebe wants to make an arithmetic tree using the numbers 421, 17, 45, and 23. She knows that the calculation 421 − 40 − 45 is an intermediate step.

  a. Draw Phoebe's arithmetic tree.

  b. Rewrite the arithmetic tree you drew as an expression with parentheses.

TRY THIS!

Section A. Arrow Language

1. Here is a record for Mr. Kamarov's bank account.

Date	Deposit	Withdrawal	Total
10/15			$210.24
10/22	$523.65	$140.00	
10/29	$75.00	$40.00	

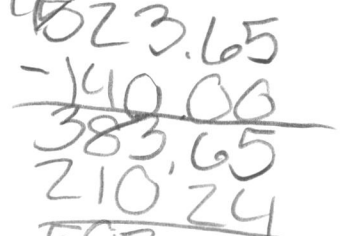

  a. Find the totals for October 22 and October 29.

  b. Write arrow strings to show how you found the totals in part **a**.

  c. When does Mr. Kamarov first have at least $600 in his account?

2. Find the results for the following arrow strings:

  a. $15 \xrightarrow{-3} \underline{} \xrightarrow{+11} \underline{}$

  b. $3.7 \xrightarrow{+1.9} \underline{} \xrightarrow{+8.8} \underline{} \xrightarrow{-1.6} \underline{}$

  c. $3{,}000 \xrightarrow{-1{,}520} \underline{} \xrightarrow{-600} \underline{} \xrightarrow{+5{,}200} \underline{}$

Section B. Smart Calculations

1. Below are some shopping problems. For each, write an arrow string to show the change that the cashier gives to the customer. Be sure to use the small-coins-and-bills-first method. Then write another arrow string that has only one arrow to show the total change.

  a. A customer gives $20.00 for a $9.59 purchase.

  b. A customer gives $5.00 for a $2.26 purchase.

  c. A customer gives $16.00 for a $15.64 purchase.

2. Rewrite the following arrow strings so that each has only one arrow:

  a. $750 \xrightarrow{+35} \underline{} \xrightarrow{+40} \underline{}$

  b. $63 \xrightarrow{-3} \underline{} \xrightarrow{+50} \underline{}$

  c. $439 \xrightarrow{+1} \underline{} \xrightarrow{-20} \underline{}$

3. Rewrite each of the following arrow strings with a new string that will make the computation easier. Explain either why your new string makes the computation easier or why this is not possible.

  a. $74 \xrightarrow{+66} \underline{}$

  b. $231 \xrightarrow{-58} \underline{}$

  c. $459 \xrightarrow{+27} \underline{}$

Try This!

Section C. Formulas

Clarinda has a personal computer at home, and she subscribes to Tech Net for Internet access. Tech Net charges $15 per month for access plus $2 per hour of usage. For example, if Clarinda is connected to the Internet for a total of three hours one month, she pays $15 plus three times $2, or $21, for the month.

1. Which of these strings gives the correct cost for Internet service through Tech Net? Explain your answer.

   a. $15 $\xrightarrow{+\ \$2}$ ___ $\xrightarrow{\times \text{ number of hours}}$ total cost

   b. number of hours $\xrightarrow{+\ \$15}$ ___ $\xrightarrow{\times\ \$2}$ total cost

   c. number of hours $\xrightarrow{\times\ \$2}$ ___ $\xrightarrow{+\ \$15}$ total cost

2. How much does it cost Clarinda for the following amounts of monthly usage:

   a. 5 hours

   b. 20 hours

   c. $6\frac{1}{2}$ hours

   10 × 2 = 20 + 15 = 35

Another company, Online Time, charges only $10 per month, but $3 per hour.

3. Write an arrow string that can be used to find the cost of Internet access through Online Time.

4. If Clarinda uses the Internet approximately 10 hours a month, which company should she use—Tech Net or Online Time?

   10 × 3 = 30 + 10 = 40

Mathematics in Context • Expressions and Formulas

Expressions and Formulas

Carlos works at a plant store that sells flower pots. One type of flower pot has a rim height of 4 centimeters and a hold height of 16 centimeters.

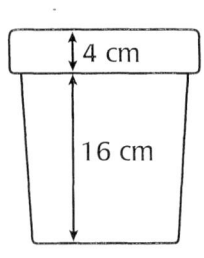

5. How tall is a stack of two of these pots? three of these pots?

6. Write a formula using arrow language that can be used to find the height of a stack if you know the number of pots.

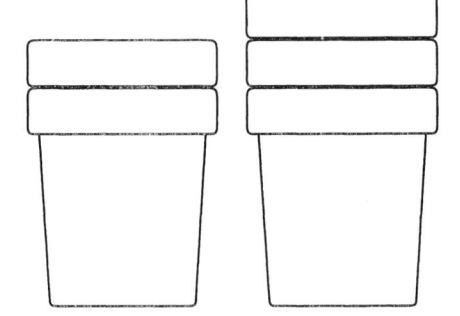

7. Carlos has to stack these pots on a shelf that is 45 centimeters high. How many can be placed in a stack this high? Explain your answer.

8. Compare the following pairs of arrow strings and decide whether they provide the same or different results:

   a. input $\xrightarrow{\times 8}$ ____ $\xrightarrow{\div 2}$ output
      input $\xrightarrow{\div 2}$ ____ $\xrightarrow{\times 8}$ output

   b. input $\xrightarrow{+ 5}$ ____ $\xrightarrow{\times 3}$ output
      input $\xrightarrow{\times 3}$ ____ $\xrightarrow{+ 5}$ output

   c. input $\xrightarrow{\div 2}$ ____ $\xrightarrow{+ 1}$ ____ $\xrightarrow{+ 6}$ output
      input $\xrightarrow{\div 2}$ ____ $\xrightarrow{+ 6}$ ____ $\xrightarrow{+ 1}$ output

Try This!

Section D. Reverse Operations

Ravi lives in Bellingham, Washington. He travels to Vancouver, Canada, quite frequently. When Ravi is in Canada, he uses the following rule to estimate prices in U.S. dollars:

number of Canadian dollars $\xrightarrow{\div 4}$ ___ $\xrightarrow{\times 3}$ number of U.S. dollars

1. Using Ravi's formula, estimate U.S. prices for the following Canadian prices:
   a. a hamburger for $2 Canadian
   b. a T-shirt for $18 Canadian
   c. a movie for $8 Canadian
   d. a pair of shoes for $45 Canadian

2. Write a formula that Ravi can use for converting U.S. dollars to Canadian dollars.

3. Write the reverse string for each of the following strings:
   a. input $\xrightarrow{-1}$ ___ $\xrightarrow{\times 2.5}$ ___ $\xrightarrow{+4}$ output
   b. input $\xrightarrow{+6}$ ___ $\xrightarrow{-2}$ ___ $\xrightarrow{\div 5}$ output

4. Find the input for each of the following strings:
   a. input $\xrightarrow{+10}$ ___ $\xrightarrow{\times 2}$ ___ $\xrightarrow{+3}$ 9
   b. input $\xrightarrow{\times 4}$ ___ $\xrightarrow{-5}$ ___ $\xrightarrow{\div 3}$ ___ $\xrightarrow{+1}$ 10

Section E. Tables

1. In your notebook, copy and complete the following tables.

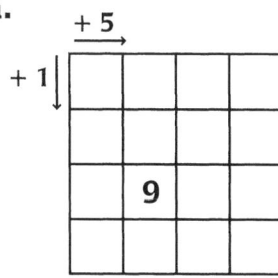

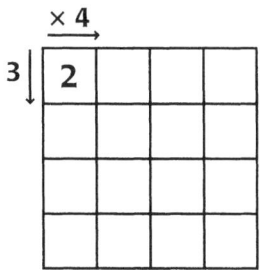

   a. (4×4 table with +5 across top, +1 down left side; 9 in third row, second column)

   b. (4×4 table with ×4 across top, ×3 down left side; 2 in second row, first column)

2. Which operations fit with the arrows marked **a** through **d** inside the following table?

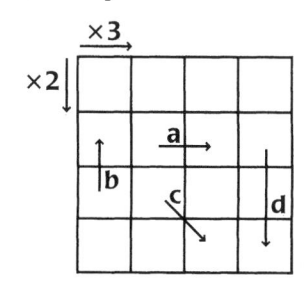

   (4×4 table with ×3 across top, ×2 down left side; internal arrows labeled a, b, c, d)

Mathematics in Context • Expressions and Formulas   57

Expressions and Formulas

3. Look carefully at the following tables for women's and men's U.S., U.K., and European shoe sizes.

Women's Shoe Sizes

U.S.	U.K.	Europe
5	$3\frac{1}{2}$	36
$5\frac{1}{2}$	4	$36\frac{1}{2}$
6	$4\frac{1}{2}$	37
$6\frac{1}{2}$	5	$37\frac{1}{2}$
7	$5\frac{1}{2}$	38
$7\frac{1}{2}$	6	$38\frac{1}{2}$
8	$6\frac{1}{2}$	39
$8\frac{1}{2}$	7	$39\frac{1}{2}$
9	$7\frac{1}{2}$	40

Men's Shoe Sizes

U.S.	U.K.	Europe
6	5	38
7	6	$39\frac{1}{2}$
8	7	41
9	8	42
10	9	43
11	10	$44\frac{1}{2}$
12	11	46
13	12	47
14	13	48

Source: Data from *How Many, How Long, How Far, How Much* (The Stonesong Press, Inc., 1996).

a. Make a list of all of the regularities you can find in the tables above.

b. If possible, write a horizontal and a vertical arrow string that could be used to generate the table for women's shoe sizes. If this is not possible, explain why.

c. Do the same as in part **b** for men's shoe sizes.

Section F. Order of Operations

1. In your notebook, copy and complete the following arithmetic trees:

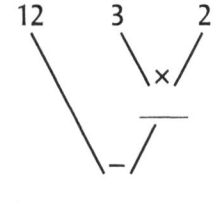

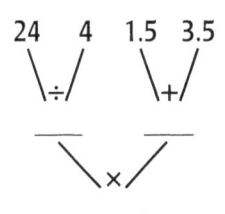

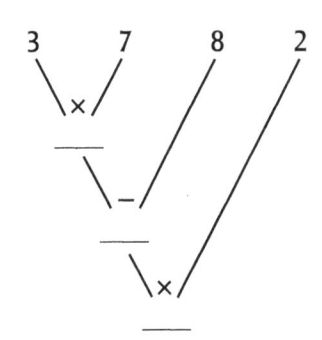

a. 12 3 2

b. 24 4 1.5 3.5

c. 3 7 8 2

2. Draw an arithmetic tree and find the answer for the following calculations:

a. $10 + 1.5 \times 6$ **b.** $(10 + 1.5) \times 6$ **c.** $15 \div (2 \times 2 + 1)$

3. Suzanne had to go to the veterinarian because her cat needed dental surgery. (Her cat never brushed its teeth!) Before the surgery, the veterinarian gave Suzanne the following estimate for the cost: $55 for anesthesia, $30 total for teeth cleaning, $18 per tooth pulled, $75 per hour of surgery, plus the cost of medicine. Make an arithmetic tree that could be used to find the total cost of Suzanne's bill from the veterinarian. Use words in your arithmetic tree where necessary.

CREDITS

Cover
Design by Ralph Paquet/Encyclopædia Britannica, Inc.

Collage by Koorosh Jamalpur/KJ Graphics.

Title Page
Brent Cardillo/Encyclopædia Britannica, Inc.

Illustrations
1–6 Phil Geib/Encyclopædia Britannica, Inc.; **7, 9** Paul Tucker/Encyclopædia Britannica, Inc.; **10** Brent Cardillo/Encyclopædia Britannica, Inc.; **12–13** Phil Geib/Encyclopædia Britannica, Inc.; **15–16** Paul Tucker/Encyclopædia Britannica, Inc.; **21–24, 27–31** Phil Geib/Encyclopædia Britannica, Inc.; **32–33** Paul Tucker/Encyclopædia Britannica, Inc.; **34 (bottom)** Paul Tucker and Tom Zielinski/Encyclopædia Britannica, Inc.; **34 (top), 35, 37–39** Phil Geib/Encyclopædia Britannica, Inc.; **41** Paul Tucker/Encyclopædia Britannica, Inc.; **45** Phil Geib/Encyclopædia Britannica, Inc.; **49–50** Paul Tucker/Encyclopædia Britannica, Inc.

Photographs
17 © Ezz Westphal/Encyclopædia Britannica, Inc.

Encyclopædia Britannica, the thistle logo, and *Mathematics in Context* are registered trademarks of Encyclopædia Britannica, Inc. Other brand and product names are the trademarks or registered trademarks of their respective owners.